Mukesh Yadav, Vikas Kumar and Nirmala Sehrawat

**Industrial Biotechnology**

## Also of interest

*Chemical Product Technology*
Murzin, 2018
ISBN 978-3-11-047531-9
e-ISBN 978-3-11-047552-4

*Product and Process Design.*
*Driving Innovation*
Harmsen, De Haan, Swinkels, 2018
ISBN 978-3-11-046772-7
e-ISBN 978-3-11-046774-1

*Environmental Engineering.*
*Basic Principles*
Tomašić, Zelić, 2018
ISBN 978-3-11-046801-4
e-ISBN 978-3-11-046803-8

*Membrane Engineering.*
Drioli, Giorno, Macedonio (Eds.), 2018
ISBN 978-3-11-028140-8
e-ISBN 978-3-11-028139-2

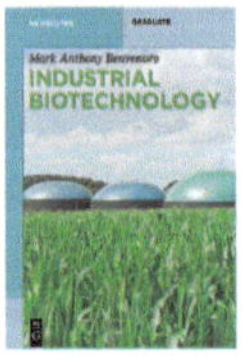


*Industrial Biotechnology.*
Benvenuto, 2019
ISBN 978-3-11-053639-3
e-ISBN 978-3-11-053662-1

# Industrial Biotechnology

Plant Systems, Resources and Products

Edited by
Mukesh Yadav, Vikas Kumar and Nirmala Sehrawat

DE GRUYTER

**Editors**
Dr. Mukesh Yadav
Maharishi Markandeshwar (Deemed to be University)
Department of Biotechnology
Engineering Block-II
Mullana-Ambala
Haryana-133207
India
mukeshyadav7@gmail.com

Dr. Vikas Kumar
Maharishi Markandeshwar (Deemed to be University)
Department of Biotechnology
Engineering Block-II
Mullana-Ambala
Haryana-133207
India
vmeashi@gmail.com

Dr. Nirmala Sehrawat
Maharishi Markandeshwar (Deemed to be University)
Department of Biotechnology
Engineering Block-II
Mullana-Ambala
Haryana-133207
India
nirmalasehrawat@gmail.com

ISBN 978-3-11-056330-6
e-ISBN (PDF) 978-3-11-056333-7
e-ISBN (EPUB) 978-3-11-056354-2

**Library of Congress Control Number: 2019936646**

**Bibliographic information published by the Deutsche Nationalbibliothek**
The Deutsche Nationalbibliothek lists this publication in the Deutsche Nationalbibliografie; detailed bibliographic data are available on the Internet at http://dnb.dnb.de.


Typesetting: Integra Software Services Pvt. Ltd.
Printing and binding: CPI books GmbH, Leck
Cover image: Reptile8488/iStock / Getty Images Plus

www.degruyter.com

MIX
Papier aus verantwortungsvollen Quellen
FSC www.fsc.org
FSC® C083411

# Preface

Industrial biotechnology is one of the most promising approaches to fulfill the ever-increasing demand of the world in terms of health, food, nutrition and other aspects of life. It offers new ways to reduce costs and create new products while protecting the environment by preventing pollution and conserving resources. The innovations in biotechnology help us to produce novel biological products. Industrial biotechnology includes use of microorganisms, plants, animal cells and enzymes to produce various products or metabolites for different industries, including food, pharmaceuticals, biodegradable plastic, biochemical industry, agricultural industry, health and diagnostics industry, and bio-fuel industry. The plant systems and resources have enormous prospective benefits to mankind. The increasing demand and continuous growth in population and undesirable climatic changes leads to decline and loss of plant resources. Therefore, there is an urgent need to develop advanced techniques for sustainable utilization of plant resources. Currently, plant biotechnology has been developed as a new age of science and technology. It is a powerful approach for the development of new plant traits and varieties. Such new varieties must be produced on a large scale to achieve commercial success as well as to satisfy the demand. Currently, the major focus of plant biotechnology is on production of secondary metabolites, valuable plant genetics improvements, germplasm conservation and production of disease-free plant verities on large scale. Plant biotechnology also works to provide the cost-effective, easily available and environment-friendly substrates in microbial biotechnology to produce microbial metabolites.

This book covers the scope and relevance of plant systems and resources for the production of industrially important products or metabolites of plant origin and microbial origin as well. The plant or agricultural waste material can be used as substrate for microbial biomass and various product formations. The waste from agro-food industry can be of economic value and may be used as source of valuable metabolites. In case of plant itself, the medicinal plants are valuable sources of herbal products, and they are disappearing at a high speed. Moreover, secondary metabolites of plant origin are of biological origin, better in terms of cost, safe to consume and have more demand as compared to synthetic compounds. Plants can be transformed genetically to either introduce new traits or alter the existing traits. Higher production of metabolites, edible vaccines, chloroplast engineering, plant tissue culture and plant pigments are also important aspects of plant biotechnology in terms of industrial point of view. This book reviews the research, concise and collective information, global trends, and developments and prospects for the strategies and methodologies concerning the sustainable use of plant resources to support the expanding plan based and other dependent industries. Transgenic plants can be the cost-effective resource to produce industrially important enzymes. Further, plants provide inexpensive

https://doi.org/10.1515/9783110563337-201

production platforms for pharmaceuticals and nutraceuticals. The concept of plant-based biosensors and their practical approach have also been included in the book. We hope that this book will provide an important insight regarding the importance of plant systems and resources for the industrial side of biotechnology. More elaborative research on methodology and commercial approach is required for sustainable use of plant resources and also for conservation of plant diversity.

# Contents

# List of contributing authors

**Chapter 1**

**Ram Sarup Singh**
Carbohydrate and Protein Biotechnology Laboratory
Department of Biotechnology
Punjabi University
Patiala 147 002
Punjab, India
rssingh11@lycos.com

**Taranjeet Singh**
Carbohydrate and Protein Biotechnology Laboratory
Department of Biotechnology
Punjabi University
Patiala 147 002
Punjab, India
taranjeetpup@gmail.com

**Chapter 2**

**Balwinder Singh Sooch**
Department of Biotechnology
Punjabi University
Patiala-147 002
Punjab, India
soochb@pbi.ac.in

**Yogita Lugani**
Department of Biotechnology
Punjabi University
Patiala-147 002
Punjab, India
yogita_lugani@yahoo.com

**Ram Sarup Singh**
Carbohydrate and Protein Biotechnology Laboratory
Department of Biotechnology
Punjabi University
Patiala 147 002
Punjab, India
rssingh11@lycos.com

**Chapter 3**

**Amit Kumar**
Department of Biotechnology
College of Natural and Computational Sciences
Debre Markos University
Debre Markos, Ethiopia
baliyaniitr@gmail.com

**Diwakar Aggarwal**
Department of Biotechnology
Maharishi Markandeshwar (Deemed to be University)
Mullana-Ambala-133207
Haryana, India
diwakaraggarwal@yahoo.co.in

**Mukesh Yadav**
Department of Biotechnology
Maharishi Markandeshwar (Deemed to be University)
Mullana, Ambala-133207
Haryana, India
mukeshyadav7@gmail.com

**Pawan Kumar**
Department of Biotechnology
Maharishi Markandeshwar (Deemed to be University)
Mullana, Ambala 133207
Haryana, India
pwn.verma2008@gmail.com

**Vikas Kumar**
Department of Biotechnology
Maharishi Markandeshwar (Deemed to be University)
Mullana (Ambala) 133207
Haryana, India
vmeashi@gmail.com

https://doi.org/10.1515/9783110563337-202

## Chapter 4

**Ashish Kumar Singh**
Department of Biotechnology
Punjabi University, Patiala, India And
Department of Biotechnology
Multani Mal Modi College, Patiala, 147001
India
ashish_rs@pbi.ac.in

**Neelam Verma**
Department of Biotechnology
Punjabi University, Patiala: 147001
India
verma.neelam2@gmail.com

## Chapter 5

**Pawan Kumar**
Department of Biotechnology
Maharishi Markandeshwar (Deemed to be University)
Mullana, Ambala 133207
Haryana, India
pwn.verma2008@gmail.com

**Pooja Sharma**
Department of Biotechnology
Maharishi Markandeshwar (Deemed to be University)
Mullana, Ambala 133207
Haryana, India
pooja0029@gmail.com

**Vikas Kumar**
Department of Biotechnology
Maharishi Markandeshwar (Deemed to be University)
Mullana, Ambala 133207
Haryana, India
vmeashi@gmail.com

**Amit Kumar**
Department of Biotechnology
Maharishi Markandeshwar (Deemed to be University)
Mullana, Ambala 133207
Haryana, India
amitvashishth659@gmail.com

**Amit Kumar**
Department of Biotechnology
College of Natural and Computational Sciences
Debre Markos University
Debre Markos, Ethiopia
baliyaniitr@gmail.com

**Anil Kumar Sharma**
Department of Biotechnology
Maharishi Markandeshwar (Deemed to be University)
Mullana, Ambala
Haryana, India
anibiotech18@gmail.com

## Chapter 6

**Navneet Kaur**
Institute of Microbial Technology (IMTech)
Chandigarh 160036, India
navi110188@gmail.com

**Vikas Kumar**
Department of Biotechnology
Maharishi Markandeshwar (Deemed to be University)
Mullana (Ambala) 133207
Haryana, India
vmeashi@gmail.com

**Parveen Rishi**
Department of Microbiology
Panjab University
Chandigarh 160014, India
rishipraveen@yahoo.com

**Nirmala Sehrawat**
Department of Biotechnology
Maharishi Markandeshwar (Deemed to be University)
Mullana-Ambala 133207
Haryana, India
nirmalasehrawat@gmail.com

**Rahul Dilawari**
Institute of Microbial Technology (IMTech)
Chandigarh 160036, India
rahuldilawari1@gmail.com

**Pawan Kumar**
Department of Biotechnology
Maharishi Markandeshwar (Deemed to be University)
Mullana (Ambala) 133207
Haryana, India
pwn.verma2008@gmail.com

**Neeraj Kumar Aggarwal**
Department of Microbiology
Kurukshetra University
Kurukshetra 136119
Haryana, India
neerajkuk26@gmail.com

**Chapter 7**

**Diwakar Aggarwal**
Department of Biotechnology
Maharishi Markandeshwar (Deemed to be University)
Mullana-Ambala-133207
Haryana, India
diwakaraggarwal@yahoo.co.in

**Anil Kumar**
TIFAC-Center of Relevance and Excellence in Agro and Industrial Biotechnology
Thapar Institute of Engineering and Technology
Patiala-147001
Punjab, India
adatta@thapar.edu

**Amit Kumar**
Department of Biotechnology
College of Natural and Computational Science
Debre Markos University
Debre Markos, Ethiopia
baliyaniitr@gmail.com

**Chapter 8**

**Vimala Devi Sadanandam**
ICAR-National Bureau of Plant Genetic Resources
New Delhi-110 012
India
vimals123@gmail.com

**Pravas Ranjan Kole**
Protection of Plant Varieties and Farmers' Rights Authority
New Delhi-110 012
India
prkole1979@gmail.com

**Rawi Gowthami**
ICAR-National Bureau of Plant Genetic Resources
New Delhi-110 012
India
gowthamir111@gmail.com

**Nirmala Sehrawat**
Department of Biotechnology
Maharishi Markandeshwar (Deemed To Be University)
Mullana, Ambala
Haryana 133207
India
nirmalasehrawat@gmail.com

**Chapter 9**

**Aman Verma**
ICAR-Directorate of Groundnut Research
Junagarh Gujarat 362001, India
verma.aman1980@gmail.com

**Chapter 10**

**Ram Kumar Pundir**
Department of Biotechnology
Ambala College of Engineering and Applied Research
Devsthali, PO Sambhlakha
Ambala-133101
Haryana, India
drramkpundir@gmail.com

**Pranay Jain**
Department of Biotechnology
University Institute of Engineering and Technology
Kurukshetra University
Kurukshetra-136119
Haryana, India
drpranayjain@gmail.com

**Satish Kumar**
Department of Biotechnology
Ambala College of Engineering and Applied Research
Devsthali, PO Sambhlakha
Ambala-133101
Haryana, India
satishrana.biotech@gmail.com

**Mukesh Yadav**
Department of Biotechnology
Maharishi Markandeshwar (Deemed to be University)
Mullana, Ambala-133207
Haryana, India
mukeshyadav7@gmail.com

**Rajesh Kumar**
Department of Agriculture Botany
Gochar Mahavidhyalaya
Rampur Maniharan
Saharanpur-247451
Uttar Pradesh, India
raj_micro78@gmail.com

# Editors' Biographies

**Dr Mukesh Yadav** is Assistant Professor at the Department of Biotechnology, Maharishi Markandeshwar (Deemed to be University), Mullana (Ambala), India. His research areas include microbial enzymes, microbial biotechnology and microbial metabolites of industrial importance. He has published research and review articles in various reputed national and international journals. He is lifetime member of several professional scientific societies and contributing to the development of society.

**Dr Vikas Kumar** is Assistant Professor at Maharishi Markandeshwar (Deemed to be University), Mullana (Ambala), India, and has more than four years of experience in teaching. He received his doctorate in Microbiology from Kurukshetra University, Kurukshetra, Haryana, India. He has guided several MSc, MTech and PhD students in their research work. He has published a book and more than 40 research papers and reviews in journals of national and international repute. He also serves as an associate editor and reviewer of various journals. As a researcher, his research interest includes areas like development of biological control agents, plant pathology and antimicrobial activity of natural and chemical compounds. Currently, he is working on a project which researches on role of endophytic fungi for improving the plant growth.

**Dr Nirmala Sehrawat** is Assistant Professor at the Department of Biotechnology, Maharishi Markandeshwar (Deemed to be University), Mullana (Ambala), India. With more than nine years of teaching and research experience, she has published research and review articles in various reputed journals. Currently, she is working on a project on food grain legumes and their industrial aspects. She is an active member of several professional and scientific societies. She has been serving as an editorial board member for several national and international journals.

https://doi.org/10.1515/9783110563337-203

R. S. Singh and Taranjeet Singh

# 1 Inulinase and pullulanase production from agro-industrial residues

**Abstract:** Inulinases and pullulanases have attracted the global attention of researchers owing to their vast applications in food industries. Inulinases hydrolyze the glycosidic linkages of inulin to produce fructooligosaccharides and high fructose syrup, whereas pullulanases hydrolyze the α-1,4 and α-1,6 glycosidic linkages in pullulan, starch and amylopectin to yield maltotriose, resistant starch, panose, isopanose, etc. Commercially, pullulanases are used in the starch liquefication and saccharification process, which reduces the use of glucoamylases by 50% in the industrial starch conversion process. Cost of substrate is an important factor in bioprocess economics. Therefore, many low-cost fermentation processes have been developed for the production of inulinases and pullulanases using various agro-industrial wastes. This chapter focuses on the production of inulinases and pullulanases in both the solid state and submerged fermentation (SmF) using agro-industrial residues, their purification, characterization as well as their applications.

**Keywords:** Inulinase, pullulanases, glucoamylases, liquefication, saccharification

## 1.1 Introduction

Inulinases, are an important class of industrial enzymes belonging to the glycoside hydrolase (GH) family 32, are characterized on the basis of their action pattern on inulin. Although inulinases have been reported from a variety of plants, animals and microorganisms, they are present in very minute quantities in plants and animals, thereby making microorganisms the best source of inulinases. Among the microorganisms, *Aspergillus* sp., *Bacillus* sp., *Kluyveromyces* sp., *Penicillium* sp., *Pseudomonas* sp. and *Streptomyces* sp. are quite efficient inulinase producers [1]. There are several advantages of microbial inulinases, such as they offer easy cultivation and genetic manipulation, high production yield, rapid multiplication, and considerable variability in biophysical and biochemical characteristics, and they are mostly inducible and extracellular in nature [2, 3]. Inulinases from fungal sources can be produced at lower substrate concentrations and have low pH and higher temperature stability, making them industrially more advantageous over other microbial sources. Inulinases have been produced in both solid state and submerged fermentations (SmFs) using a wide range of substrates and microorganisms [2, 3].

https://doi.org/10.1515/9783110563337-001

On the basis of their mode of action, inulinases are classified as exoinulinases (EC 3.2.1.80) and endoinulinases (3.2.1.7). Exoinulinases perform sequential degradation from the non-reducing end of the inulin, releasing fructose with a molecule of glucose, whereas endoinulinases act arbitrarily on the internal β-2,1-glycosidic linkages of inulin to produce fructooligosaccharides of varied chain lengths. The exoacting and endoacting natures of inulinases mainly depend on the source of the enzyme. Inulinases from fungal sources are generally exoacting. However, in some cases, fungal strains like *Aspergillus* sp., *Chrysosporium pannorum* and *Penicillium rugulosum,* and the yeast strain *Cryptococcus aureus* produce both exo- and endoinulinases [1, 2]. Although most of the microorganisms produce extracellular inulinases, in some strains the enzyme location is intracellular [4, 5], and yet some other microbial species produce both extra- and intracellular inulinases [6]. Inulinases are used for the production of high fructose syrup, fructooligosaccharides, bioethanol, single-cell oil, single-cell proteins and other beneficial bio-products, and hence they play a predominant role in various food and pharmaceutical industries [7–10].

Pullulanase-debranching enzymes belong to the GH 13 family, which hydrolyzes the glycosidic linkages in pullulan and other branched polysaccharides. Pullulanases degrade pullulan, β-limit dextrin and amylopectin, and the detailed mechanisms of substrate degradation and the final end products are different in each case. Based on substrate specificity and catalytic reaction end products, pullulan-hydrolyzing enzymes have been classified into three types: pullulanases (types I and II), pullulan hydrolases (types I, II and III) and glucoamylases. Pullulanases (types I and II) hydrolyze α-1,6 linkages in pullulan and starch to produce maltotriose. Pullulan hydrolases type I (neopullulanase) and type II (isopullulanase) break the α-1,4 linkages of pullulan to release panose and isopanose, whereas pullulan hydrolase type III breaks both α-1,4 and α-1,6 linkages in pullulan to release maltotriose, maltose and pentose. However, glucoamylase hydrolyzes α-1,4 linkages from the non-reducing end of pullulan to yield glucose as a product.

## 1.2 Microbial sources of inulinases

Microorganisms from various microbial groups like filamentous fungi, bacteria and yeast strains have been successfully used for inulinase production (Figure 1.1). Generally, filamentous fungi are preferred over bacterial and yeast strains because of their ability to grow at low pH and high temperatures, which reduce the chances of contamination and secondary product formation. Moreover, inulinase-producing fungi can also be cultivated at low-value substrates, thus making the fermentation process more cost-effective. Among the filamentous fungi, *Aspergillus* sp., *Penicillium* sp. and *Cladosporium* sp. have been reported to be prolific inulinase producers.

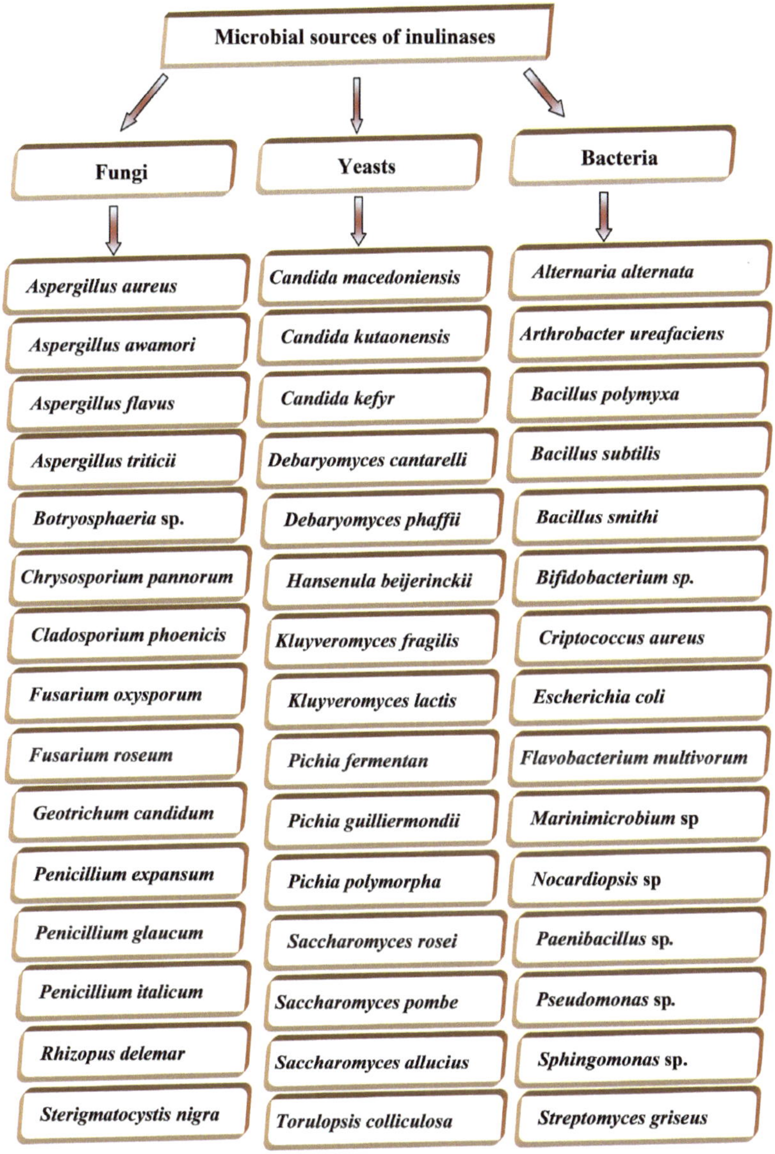


Figure 1.1: Inulinase-producing microorganisms.

*Aspergillius niger, A. niveus, A. ficuum, A. tamari, A. terreus* and *A. tubingensis* are potent inulinase-producing aspergilli, whereas *Penicillium subrubescens, P. expansum, P. purpurogenum, P. rugulosum* and *P. trzebinskii* have been reported as potent inulinase-producing penicilli. Moreover, among the various species of *Cladosporium, C. cladosporioides* and *C. phoenici* are very effective inulinase producers [2].

Bacterial strains like *Bacillus* sp., *Streptomyces* sp., *Xanthomonas* sp. and *Clostridium* sp. are used for inulinase production because of their ability to survive under unfavorable conditions like high temperature, alkalinity, acidity and salinity. Therefore, isolation of bacterial strains producing high quantities of thermostable inulinases is in great demand for industrial applications. Recently, bacterial strains like *Bacillus* sp. [11], *B. safensis* [12], *Acinetobacter baumannii* [13] and *B. centrosporus* [14] have been reported as new inulinase producers. Some extremophilic bacteria like *Streptomyces* sp., *Bacillus* sp., *Clostridium thermoautotrophicum, Xanthomonas oryzae* and *Sphingobacterium* sp. have also been reported to produce characteristically distinguishable inulinases [3].

Owing to their ease of handling and fast growth rates, yeast strains are being preferred over fungal strains for inulinase production. Among the different yeast strains, *Kluyveromyces* sp., *Pichia* sp. and *Candida* sp. are potent inulinase producers. *Kluyveromyces marxianus* is a well-known inulinase producer, whereas some of the marine-derived yeasts, such as *Meyerozyma guilliermondii* [15], *Cryptococcus aureus* [16] and *Zygosaccharomyces cerevisiae* [17], are efficient inulinase producers. Recently, a few yeast strains like *Zygosaccharomyces, Gordonia, Hanseniaspora, Torulosopra, Saccharomyces, Metschnikowia* and *Lachancea*have have also been screened for inulinase production [18].

## 1.3 Production of inulinases from agro-industrial residues

A wide range of microorganisms from filamentous fungi, bacteria, yeast and actinomycetes have been reported for inulinase production in solid-state fermentation (SSF) using agro-industrial residues as the substrate [2, 3]. The substrate and the type of microorganisms used for the production of inulinases play a vital role in the localization of the enzyme, its mode of action as well as its yield.

### 1.3.1 Submerged fermentation (SmF)

SmF is one of the most efficacious and finest techniques used for inulinase production at the industrial scale. It has been practiced since the past few centuries due to its numerous advantages like ease of process control, sterilization process, media homogeneity and appropriate heat transfer system. [19]. Although various substrates have been used for the production of inulin hydrolytic enzymes, inulin and inulin-rich plant extracts are mainly used in SmF. Raw inulin extracted from tubers of *Asparagus offcinalis* [20], *Asparagus racemosus* [21] and dahlia [22] have been used as a substrate for inulinase production in shake-flask fermentations. The degree of polymerization (DP) of inulin also plays a pivotal role in inulinase production. In a comparative study, inulinase production by *A. niger* was observed using raw and dried garlic inulin as a substrate [23]. It has been reported that dried garlic produces lower enzyme yield than raw garlic. This is because fresh raw garlic has a higher inulin content than dried garlic, where inulin may get degraded because of storage and seasonal changes.

There are various key factors that can result in successful SmF like carbon and nitrogen sources, pH, temperature, aeration and agitation. Inulin can be either used in the pure form or mixed with other natural substrates for the production of inulinases (Table 1.1). It acts as an inducer as well as a sole carbon source for different classes of microbes. Other substrates like glucose, sucrose, lactose, maltose and fructan can also be used for inulinase production [1]. Angel *et al.* [24] formulated a medium to study the effects of different carbon sources – inulin, sucrose, xylose, fructose and glucose on inulinase production from *Bacillus* sp., *Pseudomonas* sp., *Lactobacillus* sp. and *Achromobacter* sp. Although inulin is considered the most suitable carbon source for inulinase production from all the bacterial strains under SmF, modified substrates such as caproyl and cholesteryl derivatives of dahlia inulin and octadecanoylsucrose derivatives have also been used as a carbon source for inulinase production [25, 26]. Nitrogen source is the second important medium constituent used for enhancing inulinase production. Various complex nitrogen sources (peptone, beef extract, yeast extract, meat extract, corn steep liquor, urea, etc.) and inorganic nitrogen sources ($NH_4Cl$, $(NH_4)_2SO_4$, $(NH_4)_2H_2PO_4$, $(NH_4)_2HPO_4$, $NaNO_3$, $KNO_3$, etc.) have been extensively used for inulinase production [10]. Complex nitrogen sources result in better inulinase production than inorganic nitrogensources [21, 27, 28]. Despite acid being usually liberated in the medium after the utilization of ammonium salts and high acidic conditions inhibit the growth and synthesis of inulinases. Ammonium salts were found necessary to induce inulinase biosynthesis from *B. polymyxa* and *B. subtilis* [29]. Organic (casein hydrolysate) and inorganic (like $(NH_4)_2SO_4$) nitrogen sources showed synergism in stimulating bacterial growth and inulinase biosynthesis [30]. Metal ions like $Mn^{2+}$, $Ca^{2+}$, $Zn^{2+}$, $Mg^{2+}$, $Fe^{2+}$ and $K^+$ had a positive influence on microbial inulinase production [1, 2]. Fungal inulinase producers are generally cultivated in a fermentation medium at the pH range of 4.5–7.0, bacterial strains at the pH range of 4.8–7.0 and yeast strains at the pH range of 4.4–6.5. The optimal temperature for bacterial and fungal strains lies in the mesophilic to thermophilic range (28–60 °C), and for yeasts strains, it lies in the mesophilic range (30–45 °C) [25]. Most of the reports describe the mesophilic range to be more suitable for inulinase production, but there are a few thermophilic inulinase producers too, including *Bacillus smithii*, *Pseudomonas* sp., *Rhizoctonia solani* and *Streptomyces* sp. [2, 3].

Table 1.1: Microbial production of inulinases in solid-state fermentation (SSF) and submerged fermentation (SmF).

| Microorganism | Substrate used | Type of fermentation | Type of inulinase | Yield (IU/mL) | Reference |
|---|---|---|---|---|---|
| *Aspergillius tritici* | *Asparagus officinalis* extract | SmF | Endoinulinase | 25.01 | [31] |
| *Aspergillus niger* | Sucrose | SmF | Exo- & endoinulinase | 3199 | [32] |
| | Banana peel | SSF | Exoinulinase | 237 | [33] |
| | Dandelion taproot extract | SmF | Endoinulinase | 55 | [34] |
| | *Helianthus annuus* root tubers & *Lactuca sativa* roots | SSF | Exoinulinase | 440 | [35] |
| | Pure inulin | SmF | Exo- & endoinulinase | 108 IU/mg[a] | [5] |
| | Pure inulin & wheat bran | SmF | Exoinulinase | 20.42 | [36] |
| *Aspergillus tamarii* | Pure inulin | SmF | Exoinulinase | 21.53 | [37] |
| | Agrowaste extract | SmF | Exoinulinase | 71.97 | [38] |
| *Aspergillus terreus* | Artichoke leaves | SSF | Exoinulinase | 21.05 | [39] |
| *Bacillus safensis* | Pure inulin | SmF | Endoinulinase | 12.56 | [40] |
| | *Asparagus racemosus* root tubers extract | SmF | Endoinulinase | 28.67 | [12] |
| *Bacillus smithii* | Pure inulin | SmF | Endoinulinase | 135.20 | [41] |
| *Brevibacillus centrosporus* | *Jerusalem artichoke* tubers extract | SmF | Exoinulinase | 6.41 | [14] |
| *Chrysosporium pannorum* | Pure inulin | SmF | Exo- & endoinulinase | 115 | [42] |
| *Cryptococcus aureus* | Pure inulin | SmF | Exo- &endoinulinase | 85 | [43] |
| | Wheat bran & rice husk | SSF | Exo- & endoinulinase | 420.90 | [44] |

Table 1.1 (continued)

| Microorganism | Substrate used | Type of fermentation | Type of inulinase | Yield (IU/mL) | Reference |
|---|---|---|---|---|---|
| *K. marxianus* | Sugarcane molasses | SmF | Exoinulinase | 1294 | [45] |
| | Dahlia tubers extract | SmF | Exoinulinase | 140.23 | [22] |
| | Dahlia tubers extract | SmF | Exoinulinase | 55.4 | [46] |
| | *Asparagus racemosus* root tubers extract | SmF | Exoinulinase | 47.3 | [21] |
| | *Asparagus officinalis* root tubers extract | SmF | Exoinulinase | 50.2 | [20] |
| *K. marxianus var. marxianus* | Pressmud | SSF | Exoinulinase | 300.50 | [47] |
| *Penicillium citrinum* | Pure inulin | SmF | Exoinulinase | 2241 | [48] |
| | Pure inulin | SmF | Exoinulinase | 20.15 | [39] |
| *Penicillium oxalicum* | Lactose | SmF | Exoinulinase | 38.52 | [49, 50] |
| *Penicillium rugulosum* | Copra waste | SSF | Exoinulinase | 239 | [51] |
| | Pure inulin | SmF | Exo- & endoinulinase | 22.60 IU/mg[a] | [52] |
| *Penicillium* sp. | Dahlia tubers extract | SmF | Endoinulinase | 3.87 | [53] |
| *Pseudomonas* sp. | Chicory roots powder | SmF | Endoinulinase | 3 | [54] |
| *Rhizopus* sp. | Pure inulin | SmF | Endoinulinase | 1.40 IU/mg[a] | [55] |
| *Rhodotorula glutinis* | Leek powder | SmF | Exoinulinase | 30.89 | [56] |
| *Saccharomyces* sp. | Wheat bran | SSF | Exoinulinase | 90.15 | [57] |
| *Streptomyces* sp. | Jerusalem artichoke powder | SmF | Endoinulinase | 1.60 | [58] |
| | Copra waste | SSF | Exoinulinase | 131 | [59] |
| *Trichoderma viride* | Jerusalem artichoke powder | SmF | Endoinulinase | 94 | [60] |

Table 1.1 (continued)

| Microorganism | Substrate used | Type of fermentation | Type of inulinase | Yield (IU/mL) | Reference |
|---|---|---|---|---|---|
| *Ulocladium atrum* | Pure inulin | SmF | Exoinulinase | 36 | [61] |
| *Xanthomonas* sp. | Chicory roots powder | SmF | Endoinulinase | 15 | [54] |
| | Dahlia tubers extract | SmF | Endoinulinase | 11 | [62] |
| *Yarrowia lipolytica* | Pure inulin | SmF | Endoinulinase | 62.85 | [63] |

[a] Specific activity of endoinulinase in crude extract

## 1.3.2 Solid-state fermentation (SSF)

Recently, SSF has become the most prominent and efficient method for inulinase production at the industrial scale. SSF has become an attractive alternative to SmF due to its numerous advantages, such as use of low-value substrates, which makes the bioprocess cost-effective; enhances product concentration; better oxygen circulation; providing physical support to growing cells; higher yield and recovery; environmental friendly nature; and reduced energy requirement and water consumption [64]. Various factors like type of substrate, moisture, water activity, pH, temperature, aeration and agitation are important effective SSF. Many agro-industrial residues and inulin-rich plant parts, such as wheat bran, rice bran, pressmud, sugarcane bagasse, chicory leaves and roots, artichoke leaves, garlic and onion peels, have been used as substrates for inulinase production in SSF using different microbial strains (Table 1.1). Both high and low moisture levels are vulnerable for inulinase production during SSF. Because yeast and fungal strains can grow at a low moisture level, they can be seamlessly used in SSF. Bacterial strains like *Staphylococcus* sp. [65], *Xanthomonas campestris* pv. *phaseoli* [66], *Streptomyces* sp. [67] and *P. rugulosum* [51] and fungal strains like *Kluyveromyces* sp. [19], *Aspergillus niger* [68], *A. parasiticus* [69] and *Penicillium oxalicum* [70] have also been used for inulinase production under SSF. Inulinase yield of 78.29 ± 0.13 U/gds was obtained from *Saccharomyces* sp. when wheat bran was used as a substrate in SSF [70]. Mostly inulinase production from fungal strains is reported in the acidic range (4.5–5.5) with the exception of *A. terreus*, which produces inulinase at pH 8.5 [71]. However, bacterial strains are normally resistant to acidic conditions and grow well in the neutral range of pH 6.5–7.5.

## 1.4 Purification and characteristics of inulinases

To determine the physicochemical characteristics of an enzyme, its purification and characterization are mandatory. Currently, various purification techniques like salt or solvent precipitation, ion-exchange chromatography, affinity chromatography, size-exclusion chromatography, hydrophobic-interaction chromatography and ultrafiltration are used for inulinase purification. Most of the inulinases are extracellular in nature; for the purification of intracellular inulinases, an additional step of cell-wall disruption is required, followed by the techniques mentioned above. The purification techniques and characteristics of some inulinases are listed in Table 1.2. Different inulinases have different hydrolytic actions (exo- or endoacting) and structural conformations depending on their source and growth conditions. Five exoinulinases (Exo-I, II, III, IV and V) and two endoinulinases (Endo-I and II) from *A. ficuum* were purified using ammonium sulfate precipitation, ion-exchange chromatography and fast protein liquid chromatography [6]. Similarly, three exoinulinases and two endoinulinases from *A. ficuum* JNSP5-06 were purified using ion-exchange chromatography and gel-filtration chromatography [72]. Recently, three active forms of inulinases (INI, INII and INIII) from *Ulocladium atrum* were purified by ammonium sulfate and ion-exchange chromatography. Inulinase (INI) was resolved into INIa, INIb and INIc by chromatographic separation with 3.43-fold purification [61]. In the case of bacterial inulinases, salt precipitation shows better results than solvent precipitation as some of the organic solvents have been reported to repress the enzyme. For salt precipitation, generally ammonium sulfate is used. In the case of *B. smithii*, ammonium sulfate and ion-exchange chromatography were used for the purification of thermostable endoinulinases [73]. Most of the bacterial inulinases are purified by ion-exchange and size-exclusion chromatography, but there are a few reports on inulinase purification by affinity chromatography. For example, recombinant inulinase from *P. polymyxa* and *Sphingomonas* sp. have been purified using $Ni^{2+}$-NTA affinity chromatography [74, 75]. Column chromatography using DEAE Sepharose and Sephacryl-200 was used for the purification of inulinase (Inu 2 and Inu 3) from *Rhiizopus oligosporus* [76]. In another case, ion-exchange and gel-exclusion chromatography were used for the purification of one endoinulinase and two exoinulinases (F2 and F3) from *Chrysosporium pannorum* [77, 78].

The conformational alternations of a biocatalyst can be understood by its molecular weight ($M_r$) and the Michaelis–Menten constant ($K_m$ and $V_{max}$). The molecular weight of fungal inulinases mainly ranges between 30 and 175 KDa and that of bacterial inulinases between 45 and 600 KDa [2, 3]. $K_m$ and $V_{max}$ of purified bacterial and fungal inulinases show considerable variations. Metal ions like $Fc^{3+}$, $Mn^{2+}$, $Cu^{2+}$, $Mg^{2+}$, $Co^{2+}$, $Na^{+}$ and $Ag^{+}$ may either stimulate or inhibit enzyme catalysis by affecting their premolecular organization [2, 3].

Table 1.2: Purification and characterization of inulinases.

| Microorganism | Purification techniques | Characteristics | | | | | | | | | | Reference |
|---|---|---|---|---|---|---|---|---|---|---|---|---|
| | | | Optimum | | Stability | | Kinetics | | Metal ions | | | |
| | | Specific activity (IU/mg) | pH | Temperature (°C) | pH | Temperature (°C) | $K_m$ (mg/mL) | $V_{max}$ (mg/min) | Activator | Inhibitor | Mr (KDa) | |
| *Paenibacillus* sp. D9 | $(NH_4)_2SO_4$ precipitation, HiTrap-QFF & Capto-MMC column chromatography | 4333 | 4.0 | 40 | 4.0–6.0 | 40 | $5.5^a$ | $476.2^b$ | $Ca^{2+}$ | $Hg^+$, $Fe^{2+}$, $Cu^{2+}$ | 58.5 | [80] |
| *Lactobacillus paracasei* | IEC on DEAE cellulose | InuB41: 159.1<br>InuLC1: 165.3 | 4.9<br>4.8 | 50<br>52 | NS<br>NS | NS<br>NS | $1.57^a$<br>$3.97^a$ | $43.7^c$<br>$66.4^c$ | NS<br>NS | NS<br>NS | 100.21<br>100.33 | [81] |
| *Aspergillus niger* | Affinity chromatography on $Ni^{2+}$-NTA | NS | 4.3 | 60 | NS | NS | $3.18^a$ | $3873.60^c$ | NS | NS | NS | [82] |
| *Kluyveromyces cicerisporus* | Affinity chromatography on $Ni^{2+}$-NTA sepharose | NS | 4.5 | 55 | 3.0–6.0 | 50 | $0.322^a$ | $4317^c$ | NS | $Cu^2$, $Co^{2+}$, $Fe^{2+}$ | 79.5 | [83] |

| | | | | | | | | | | | | |
|---|---|---|---|---|---|---|---|---|---|---|---|---|
| *Kluyveromyces marxianus* | $(NH_4)_2SO_4$ precipitation, dialysis, $Ni^{2+}$-NTA affinity chromatography, Sephacryl S-100, DEAE sepharose column | 1.38[d] | 4.5 | 55 | 4.0–5.5 | 45 | 107.40 | 1.94[e] | NS | NS | 77.2 | [84] |
| *Aspergillus niger* | $(NH_4)_2SO_4$ precipitation, dialysis, $Ni^{2+}$-NTA affinity chromatography, Sephacryl S-100, DEAE sepharose column | 9.04[d] | 4.5 | 60 | 4.0–5.5 | 50 | 63.13 | 12.83[e] | NS | NS | 66.1 | [79] |
| *Aspergillus niger* | Affinity chromatography on $Ni^{2+}$-NTA sepharose | NS | 4.3 | 53 | NS | NS | 2.86a | 2368.92[c] | NS | NS | 58 | [84] |
| *Sphingomonas* sp. | Affinity chromatography on $Ni^2$+-NTA | 73.7 | 5.5 | 55 | 4.0–7.0 | 10–70 | 10.20 | 294.10[c] | NS | $Hg^{2+}$, $Ag^{+}$, $Mn^{+}$, | 55 | [75] |
| *Rhizopus oligosporus* | IEC on DEAE Sepharose & Sephacryl S-200 CL-6B | Inu 2: 3258 Inu 3: 7238 | 5.0<br>5.0 | 50<br>60 | 4.5–6.0<br>4.5–6.0 | 60<br>70 | 0.93[f]<br>0.70 | NS<br>NS | NS<br>NS | $Ni^{2+}$, $Cu^{2+}$, $Fe^{2+}$, $Co^{2+}$, $Hg^{2+}$ | 76<br>30 | [76] |

(continued)

Table 1.2 (continued)

| Microorganism | Purification techniques | Characteristics | | | | | | | | | | Reference |
|---|---|---|---|---|---|---|---|---|---|---|---|---|
| | | | Optimum | | Stability | | Kinetics | | Metal ions | | | |
| | | Specific activity (IU/mg) | pH | Temperature (°C) | pH | Temperature (°C) | $K_m$ (mg/mL) | $V_{max}$ (mg/min) | Activator | Inhibitor | Mr (KDa) | |
| *Penicillium* sp. | Ethanol precipitation, ion-exchange chromatography | 81.7[g] | 5.0 | 50 | NS | NS | NS | NS | $Mn^{2+}$, $Ca^{2+}$ | $Hg^{2+}$ | 68 | [53] |
| *Ulocladium atrum* | $(NH_4)_2$ $SO_4$ precipitation, dialysis, ion-exchange chromatography | 2479.2 | NS | NS | NS | NS | NS | NS | NS | $Fe^{3+}$, $Mn^{2+}$, $Ba^{2+}$, $Co^{2+}$, $Mg^{2+}$ | NS | [61] |
| *Paenibacillus polymyxa* | Affinity chromatography on $Ni^{2+}$-NTA, GEC on Sephacryl S-300 | 18.50 | 6.0 | 25 | 5.5–8.5 | 20–35 | 1.72[a] | 21.69[c] | $Zn^{2+}$, $Fe^{2+}$, $Mg^{2+}$ | $Co^{2+}$, $Cu^{2+}$, $Ni^{2+}$ | 56 | [74] |
| *Streptomyces* sp. | $(NH_4)_2SO_4$ precipitation, IEC on Macro-prep DEAE, GEC on Sephacryl S-200, HIC on t-Butyl, Hydroxyapatite Bio-Gel HTP | 290.20 | 6.0 | 55 | 5.0–9.0 | 30–50 | 2.34[a] | 440[c] | $Mg^{2+}$, $Mn^{2+}$, $Co^{2+}$ | $Hg^{2+}$ | 73 | [58] |

| | | | | | | | | | | | | |
|---|---|---|---|---|---|---|---|---|---|---|---|---|
| *Candida kutaonensis* | Ultrafiltration, IEC on DEAE Sepharose & Q-Sepharose, GEC on Sephadex 200 PG | 188 | 4.5 | 50 | 3.0–6.5 | 45–60 | 2.30 | 4.80 | $Fe^{2+}$ | $Ag^{+}$, $Mn^{2+}$ | 55 | [85] |
| A. *foetidus* | $(NH_4)_2$ $SO_4$ & solvent precipitation, dialysis, GEC on Sephadex G-100, IEC on DEAE cellulose | 100.90 | 5.5 | 50 | 5.5 | NS | 3.45 | 55[c] | NS | NS | 78 | [86] |
| A. *niger* | Three-phase partitioning | 43.87 | 4.6 | 60 | 3.6–6.5 | 35–75 | 14.80 | NS | $Mg^{2+}$ | $Hg^{2+}$, $Pb^{2+}$ | 63.80 | [87] |
| A. *terreus* | Ultrafiltration, IEC on CM Sepharose & Phenyl Sepharose, GEC on Sephadex G-75 | 38.30 | 4.0 | <40 | 4.0–6.0 | 40–50 | 48.09 | 18.49[f] | $Mn^{2+}$ | $Cu^{+}$, $Ag^{+}$ | 57 | [88] |

[a] $K_m$ is expressed in mM;
[b] $V_{max}$ is expressed in IU/mg;
[c] $V_{max}$ is expressed in μmol/min/mg;
[d] Specific activity expressed in U/μmol;
[e] $V_{max}$ is expressed in μmol/mL/min;
[f] $K_m$ and $V_{max}$ values are expressed in mM/min;
[g] Specific activity expressed in nkat/mg.
[h] NS, not specified; DEAE, diethyl amminoethyl; GEC, gel-exclusion chromatography;
[i] HIC, hydrophobic-interaction chromatography; IEC, ion-exchange chromatography;
[j] QFF, Q-Sepharose Fast Flow

## 1.5 Applications of inulinases

Inulinases have tremendous applications in the food and pharmaceutical industries. Inulinases have been mainly used for the production of high fructose syrup [46–91] and fructooligosaccharides [1, 7], as well as for the production of industrially important metabolites that are generated in a bi-enzymatic process in which inulin is hydrolyzed first, followed by the conversion of the inulin hydrolysate into a major product. Simultaneous saccharification and fermentation of inulin/inulin-rich extracts by inulinases have been performed for the production of various products like ethanol [92], single-cell oil [93], single-cell proteins [94], citric acid [95], lactic acid [96], sorbitol [97], mannitol [98], pullulan [99] and 2, 3 butanediol [100].

## 1.6 Microbial sources of pullulanases

Pullulanase was first isolated from *Klebsiella pneumoniae* by Bender and Wallenfels in 1961 [101]. Pullulanases are widely distributed among various plants, animals, bacteria, yeast and fungi, which specifically hydrolyze α-1,4 and α-1,6 glycosidic linkages in pullulan, amylopectin and other related polymers. To date, many microorganisms, including mesophiles, thermophiles and hyperthermophiles, have been identified as potent pullulanase producers. Pullulanase production from various bacterial species like *Anoxybacillus* sp. [102], *Bacillus pseudofirmus* [103], *B. naganoensis* [104], *B. cereus* [105], *B. deramificans* [106], *B. flavothermus* [107], *Klebsiella variicola* [108] and *Paenibacillus barengoltzii* [109] have been reported. Among the fungal strains, *Auerobasidium pullulans* has been reported for the production of pullulanase [110] and pullulan [111–115]. Use of all these strains for the commercial production and applications of pullulanase are still low due to their low yield. Therefore, attempts have been made to overexpress pullulanase by the recombinant DNA technology using *Pichia pastoris* and *E. coli* as the expression vectors, which are being used for the heterologous expression of pullulanase due to their ease of culture, short fermentation period and high-efficiency protein synthesis. The highest yield reported in an *E.coli* expression system was 1567.90 IU/mL [116]. To meet the requirement of the industrial starch saccharification process, the saccharifying enzymes must be thermotolerant to perform well in the overall liquefaction and saccharification processes under high-temperature conditions. Therefore, globally, researchers have focused on the overexpression of genes encoding thermostable pullulanases from diverse microorganisms, including *Bacillus deramificans* [117], *Bacillus flavothermus* KWF-1 [107], *Bacillus naganoensis* [118] and *Klebsiella variicola* [119]. Furthermore, the high-temperature conditions reduce the chances of contamination, decrease energy consumption and the reaction time, and increase the product yield. Thermostable pullulanases have also been reported from *Thermotoga neapolitana* [120], *Bacillus flavothermus* [107], *Bacillus acidopullulyticus* [95], *Anoxybacillus* sp. [96], etc.

## 1.7 Types of pullulanases

Pullulanases have been classified into three types, and the mode of action of each type is different, resulting in distinct products.

### 1.7.1 Pullulanases (types I and II)

Type I pullulanase (EC3.2.1.41) typically hydrolyzes α-1,6-glycosidic linkages in pullulan and other molecules such as starch, amylopectin, glycogen and β-limit dextrin to produce maltotriose units. Thermostable type I pullulanase plays an important role in the starch industry as it withstands the high temperature of the starch saccharification process, leading to the production of α-1,4 linked glucan units, which allow more rapid and efficient conversion reactions. Among the thermostable type I pullulanase-producing microorganisms, *Bacillus acidopullulyticus* [97], *Bacillus flavocaldarius* [107], *Bacillus* sp. [121], *Thermotoga neapolitana* [120], etc. have been reported as potent pullulanase producers. Besides maltotriose, amylose and glucose are also produced by type I pullulanases with or without α-amylase and glucoamylase. Type II pullulanase (amylopullulanase; EC3.2.1.1/41) hydrolyzes α-1,6-linkages in pullulan to produce maltotriose and α-1,4-glycosidic linkages in starch to yield a mixture of glucose, maltose and maltotriose (Figure 1.2). Amylopullulanase are of great interest in the starch-processing industry as they can combine the liquefaction and saccharification processes into one step for making various sugar syrups and as an anti-staling agent in bread and detergent additives [121]. Therefore, efforts are being made to find out such aciduric and thermoduric enzymes. A thermophilic amylopullulanase-producing strain was isolated that could convert pullulan to glucose, maltose and maltotriose as well as starch to glucose and maltose [122]. Amylopullulanases are used for the structural analyses of oligo- and polysaccharides as they precisely recognize the difference between α-1,4- and α-1,6-glycosidic linkages. The amylase and pullulanase activities of type II pullulanase show the same range of pH and temperature [123].

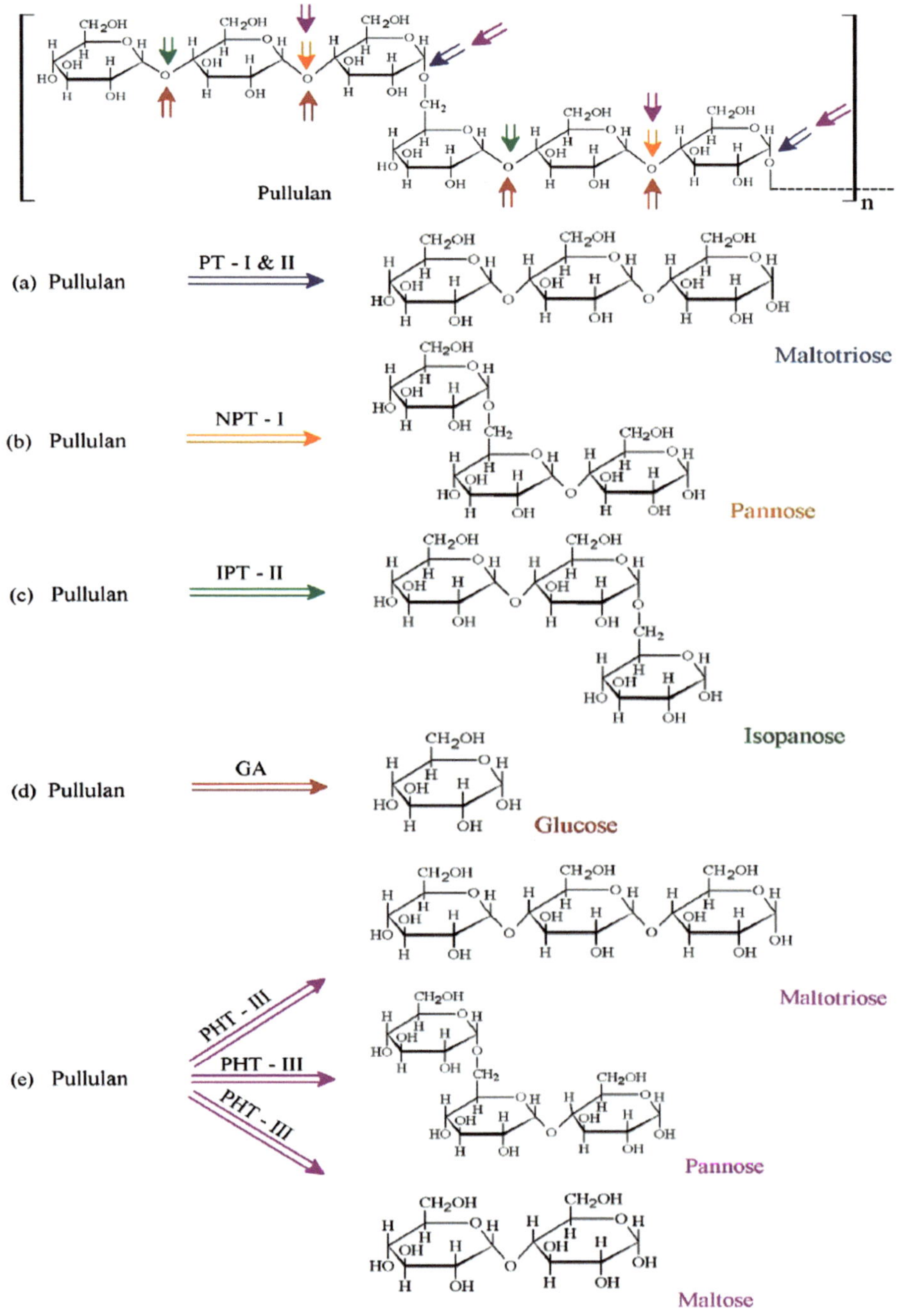


Figure 1.2: Schematic diagram showing the action patterns of pullulanases. PT-I and II, pullulanase types I and II; NPT I, neopullulanase type I; IPT-II, isopullulanase type II; GA, glucoamylase; PHT-III, pullulanase hydrolase type III.

### 1.7.2 Pullulanase hydrolases (types I, II and III)

Type I pullulanase hydrolase (neopullulanase; EC:3.2.1.135) breaks the α-1,4-glycosidic linkages in pullulan to yield panose. The other pullulan-degrading enzymes are α-amylases, which hydrolyze the α-1,4-glycosidic linkage of pullulan to produce panose. The α-amylases from *Thermoactinomyces vulgaris* [124] and *Bacillus licheniformis* [125] efficiently degrade starch than pullulan to maltotriose units and are designated as neopullulanase-α-amylase; EC: 3.2.1.1/135 [126]. Type II pullulanase hydrolase (EC: 3.2.1.57), called isopullulanase/pullulan 4-glucanohydrolase, cleaves α-1,4-glycosidic linkages of pullulan to release isopanose as the main product; at the same time, it hydrolyzes panose to yield small amounts of glucose and isomaltose. Both pullulanase hydrolases types I and II are highly active toward the degradation of cyclodextrin and cleave α-1,4-glycosidic linkages of pullulan to release panose and isopanose, respectively, whereas pullulanase hydrolase type III cleaves both α-1,4- and α-1,6-glycosidic linkages in pullulan-producing maltotriose, panose, maltose and glucose. Therefore, they are being used in combination with amylopullulanase in the starch-processing industry because the end products of both reactions are the same. Pullulanase hydrolase type III also hydrolyzes other polysaccharides like starch, amylose, amylopectin, glycogen and cyclodextrin to release maltose as the major product. Owing to its high thermostable nature, it is used in the food, chemical and pharmaceutical industries in the synthesis of non-fermentable anticarcinogenic carbohydrates and as anti-staling agents in bakery products [123].

### 1.7.3 Glucoamylases

Glucoamylases (EC: 3.2.1.3), also called exoglucanases, glucan 1,4-α-glucosidase, amyloglucosidase, γ-amylase and exo-1,4-α-glucosidase, split α-1,4-glycosidic linkages of pullulan to yield glucose. Many bacterial species like *Flavobacterium* and *Thermoanaerobacterium therosaccharolyticum* have been reported as potent glucoamylase producers [127]. The pure form of glucoamylases is used in the hydrolysis of dextrins to glucose units, which are further used in the production of high fructose syrup and ethanol. Glucoamylases can also be used in combination with type I pullulanases, which can together reduce the time of the starch saccharification process without the formation of side products, thereby increasing the process yield [128].

## 1.8 Production of pullulanase from agro-industrial residues

Only a few reports are available on pullulanase production from agro-industrial residues in the solid state and SmF. Various carbon sources including potato, sorghum, sweet potato, ragi, barley, rice, corn, wheat, tapioca flours and soluble starch have been efficiently utilized for pullulanase production from *Clostridium thermosulfurogenes* in SmF. Under optimized conditions, potato flour was found to be the most suitable carbon source for enzyme production [129]. Moreover, using low-cost substrates like potato flour instead of soluble starch reduces the production cost of pullulanases, which in turn helps in increasing the bioprocess economics. However, another study found soluble starch to be the best carbon source for biocatalyst production [130]. To meet growing industrial requirements, researchers are currently interested in thermostable pullulanases. The production of thermostable amylopullulanase in SmF has been reported from *Geobacillius thermoleovorans*. After process optimization, 840 IU/l of pullulanase was produced at the shake-flask level, whereas at the laboratory-scale bioreactor, a two-fold increase in enzyme production (1900 IU/l) was achieved with an overall reduction in fermentation time to 7–8 h. [131]. Kar et al. (2012) [132] also reported amylopullulanase production from *Streptomyces erumpens* in SmF.

Bacterial systems are also being investigated for enzyme and metabolite production in SSF. Because of its numerous advantages, SSF is being preferred over Smf for pullulanase production. SSF is a suitable and economical method because it uses simple and cheap media. Moreover, there are less chances of contamination in SSF as most of the contaminants are unable to grow in the absence of free-flowing water. Bacterial species like *Bacillus licheniformis* [133] and *Clostridium thermosulforegenes* [134–136] can efficiently produce pullulanase after process optimization in SSF.

## 1.9 Purification and characteristics of pullulanases

Biocatalysts used for industrial applications must be in the purified form and their physicochemical characteristics should be known. Purification techniques and characteristics of some pullulanases are listed in Table 1.3. Different pullulanases have different molecular masses. The molecular mass of pullulanase type I ranges from 65 to 145 kDa, whereas that of pullulanase type II lies between 45 and 250 kDa. The size of pullulanase hydrolase type I is between 58 and 83 kDa, those of pullulanase hydrolase type II from *Aspergillus niger* and *Bacillus* sp are between 59 and 200 kDa and those of pullulanase type III from *T. aggregans* and *T. kodakarensis* are 80 and 84 kDa, respectively [137].

Currently, Ni-NTA affinity chromatography is in great demand for the production of recombinant pullulanases. Various recombinant pullulanases have been purified using this technique, including *Bacillus pseudofirmus* [103], *Anoxybacillus* sp. [138] and *Thermus thermophilus* [139]. For *Klebsiella variicola* [108] and *Hermetia illucens* [140], Ni-His Trap affinity column and cobalt affinity chromatography were used for

Table 1.3: Purification and characterization of pullulanases.

| | | Characteristics | | | | | | | | | | |
|---|---|---|---|---|---|---|---|---|---|---|---|---|
| | | | Optimum | | Stability | | Kinetics | | Metal ions | | | Reference |
| Microorganism | Purification techniques | Specific activity (IU/mg) | pH | Temperature (°C) | pH | Temperature (°C) | $K_m$ (mg/mL) | $V_{max}$ (mg/min) | Activator | Inhibitor | Mr (KDa) | |
| *Anoxybacillus* sp. | Ni-NTA affinity chromatography | NS | 6.0 | 60 | 5.0–6.0 | 30–60 | 2.0 | 0.1[a] | $K^+$, $Na^+$, $Mg^{2+}$ | NS | 80.0 | [102] |
| | Ni-NTA superflow affinity chromatography | 750.0 | 6.0 | 60 | NS | NS | 16.4 | NS | NS | NS | NS | [142] |
| *Bacillus pseudofirmus* | Ni-NTA affinity chromatography | 270 | 7.0–8.0 | 45 | 5.5–9.5 | 25–35 | 0.31 | 0.17[a] | NS | $Mn^{2+}$, $Co^{2+}$, $Hg^{2+}Ni^{2+}$ | 87 | [103] |
| *Hermetia illucens* | Cobalt affinity chromatography | NS | 9.0 | 40 | 5.0–11.0 | 30 | 1.5[b] | 306.8[c] | NS | $Mn^{2+}$, $Zn^{2+}$,$Co^{2+}$ | 164 | [140] |
| *Paenibacillus barengoltzii* | Ni-IDA agarose affinity chromatography | 68.3 | 5.5 | 50 | 5.5–10.5 | 40 | 2.94 | 280.5[c] | $Co^{2+}$, $Fe^{3+}$ | $Hg^+$, $Sn^{2+}$, $Cu^{2+}$, $Ag^+$ | 75 | [109] |
| *Thermus thermophilus* | Ni-NTA affinity chromatography | 280 | 6.5 | 70 | 6.0–8.0 | 60–70 | 23.7 | 0.0031[a] | $Mn^{2+}$ | $Hg^{2+}$, $Cu^{2+}$ | 52 | [139] |

(continued)

Table 1.3 (continued)

| Microorganism | Purification techniques | Characteristics | | | | | | | | | | | Reference |
|---|---|---|---|---|---|---|---|---|---|---|---|---|---|
| | | | Optimum | | Stability | | Kinetics | | Metal ions | | | | |
| | | Specific activity (IU/mg) | pH | Temperature (°C) | pH | Temperature (°C) | $K_m$ (mg/mL) | $V_{max}$ (mg/min) | Activator | Inhibitor | Mr (KDa) | | |
| *Hypocrea jecorina* | $(NH_4)_2SO_4$ precipitation, IEC on DEAE-cellulose, GEC on Sephadex G-100 | 1.36 | 6.5 | 35–65 | NS | 40–70 | 38.4 | 3.82 | $Co^2$, $Ni^{2+}$, $Na^+$, $Mg^{2+}$ | $Ca^{2+}$ | | | 130.56 |
| [141] | | | | | | | | | | | | | |
| *Bacillus cereus* | Ni-NTA superflow affinity chromatography | 44.7 | 6.0 | 40 | 6.0–9.0 | 20–40 | 0.45 | 45.3[a] | $Ca^{2+}$ | $Ni^{2+}$, $Mn^{2+}$ | 81.4 | | [142] |
| *Klebsiella variicola* | Ni-HisTrap affinity column | NS | 5.0 | 55 | 5.0–7.5 | NS | NS | NS | NS | NS | 118 | | [108] |
| *Thermococcus kodakarensis* | Ni-NTA affinity, DEAE chromatography | NS | 5.5 | 105 | 4.0–7.0 | 85 | 2.2 | 92.3[d] | NS | $Zn^{2+}$, $Cu^{2+}$, $Fe^{2+}$ | 120 | | [143] |
| *Thermotoga neapolitana* | Ni-NTA affinity chromatography | 28.7 | 6.5 | 80 | 4.5–7.0 | NS | NS | NS | $Ca^{2+}$, $Mg^{2+}$ | $Cu^{2+}$, $Hg^+$ | 93 | | [120] |

| | | | | | | | | | | | | |
|---|---|---|---|---|---|---|---|---|---|---|---|---|
| *Geobacillus stearothermophilus* | $(NH_4)_2SO_4$ precipitation IEC on Q-Sepharose, DEAE-Sepharose, GEC on Sephadex G-100 | 967 | 5.5 | 65 | 5.0–7.0 | 60–70 | 0.48[e] | 44[a] | $Ca^{2+}$ | NS | 100 | [122] |

[a] $V_{max}$ is expressed in μmol/min;
[b] $K_m$ is expressed in mM;
[c] $V_{max}$ is expressed in μmol/min/mg;
[d] $V_{max}$ is expressed in IU/mg;
[e] $K_m$ is expressed in g/l.
[f] NS, not specified; DEAE, diethyl amminoethyl; GEC, gel-exclusion chromatography;
[g] IEC, ion-exchange chromatography.

pullulanase purification. Cobalt affinity chromatography was used for the purification of pullulanase from *Hermetia illucens*. For *Hypocrea jecorina* [141] and *Geobacillus stearothermophilus* [122], pullulanases were purified by ion-exchange chromatography on DEAE-cellulose and DEAE-sepharose, respectively, followed by gel-exclusion chromatography on Sephadex G-100.

The Michaelis–Menten constant ($K_m$) and maximal velocity ($V_{max}$) are important characteristics of an enzyme. $K_m$ and $V_{max}$ help understand the enzyme kinetics. Different pullulanases differ in their substrate specificities and hydrolysis products because of the disparity with the amino acid sequence as well as the dissimilar domain organization among the pullulanases. The $K_m$ and $V_{max}$ values of purified pullulanases from different strains vary in the range of 0.31–38.4 mg/mL and 0.17–306.8 μmol/min/mg, respectively. Metal ions like $Mn^{2+}$, $Co^{2+}$, $Hg^{2+}$, $Ni^{2}$, $Sn^{2+}$, $Cu^{2+}$, $Ag^{+}$, $Ni^{2+}$, $Na^{+}$, $Mg^{2+}$ and $Ca^{2+}$ may either enhance or decrease the enzyme activity, thereby acting as an activator or inhibitor of pullulanases (Table 1.3).

## 1.10 Application of pullulanases

Industrially, pullulanases are used for the production of glucose (30–50% glucose) and maltotriose (30–50% maltose) syrup, which are widely used in the food, pharmaceutical and beverage industries. Pullulanases are also used in starch liquefaction and saccharification processes as pullulanase type II combines both liquefaction and saccharification steps into a one-step process, thus mitigating the use of α-amylases and β-amylases in starch liquefaction and starch saccharification, respectively. Moreover, pullulanases have also been used for the production of resistant starch [144], panose and isopanose syrups [145], low-calorie beer [146], maltotriose syrup [147–149], additives in dishwashing and laundry detergents [150] and as an anti-staling agent in bakery products [151].

## References

[1] Singh RS, Singh RP. Production of fructooligosaccharides from inulin by endoinulinases and their prebiotic potential. Food Technol Biotechnol 2010, 48, 435–50.

[2] Singh RS, Chauhan K. Production, purification, characterization and applications of fungal inulinases. Current Biotechnol 2016, 5, 1–20.

[3] Singh RS, Chauhan K, Kennedy JF. A panorama of bacterial inulinases: Production, purification, characterization and industrial applications. Int J Biol Macromol 2017, 96, 312–22.

[4] Gern RM, Furlan SA, Ninow JL, Jonas R. Screening for microorganisms that produce only endo-inulinase. Appl Microbiol Biotechnol 2001, 55, 632–35.

[5] Nakamura T, Kuramori K, Zaita N, Akimoto H, Ohta K. Purification and properties of intracellular exo-and endoinlulinases from *Aspergillus niger* strain 12. Bull Fac Agr Miyazaki Univ 2001, 48, 49–58.

[6] Ettalibi M, Baratti JC. Purification, properties and comparison of invertase, exoinulinases and endoinulinases of *Aspergillus ficuum*. Appl Microbiol Biotechnol 1987, 26, 13–20.
[7] Singh RS, Singh RP, Kennedy JF. Recent insights in enzymatic synthesis of fructooligosaccharides from inulin. Int J Biol Macromol 1987, 85, 565–72.
[8] Singh RS, Chauhan K, Singh RP. Enzymatic approaches for the synthesis of high fructose syrup. In: Gahlawat S, Salar R, Siwach P, Duhan J, Kumar S, Kaur P., eds. Plant Biotechnology: Recent Advancements and Developments. Springer, Singapore, 2017, 189–211.
[9] Singh RS, Singh RP, Kennedy JF. Immobilization of yeast inulinase on chitosan beads for the hydrolysis of inulin in a batch system. Int J Biol Macromol 2017, 95, 87–93.
[10] Singh RS, Singh RP. Inulinases. In: Pandey A, Negi S, Soccol CR., eds. Current Developments in Biotechnology and Bioengineering. Elsevier Inc, Cambridge, US, 2017, 423–46.
[11] Gavrailov S, Ivanova V. Isolation and characteristics of a thermophilic *Bacillus* strain, producer of inulinase. J Biosci Biotechnol 2014, SE/ONLINE 83–94.
[12] Singh RS, Singh RP. Response surface optimization of endoinulinase production from a cost effective substrate by *Bacillus safensis* AS-08 for hydrolysis of inulin. Biocatal Agri Biotechnol 2014, 3, 365–72.
[13] Muslim SN, Ali ANM, Salman IMA, Al-Kadmy IMS, Muslim SN. Detection of the optimal conditions for inulinase productivity and activity by *Acinetobacter baumannii* isolated from agricultural rhizosphere soil. Int J Adv Chem Eng Biol Sci 2015, 2, 1–7.
[14] Zhang Y, Zhu H, Wang J, Zhou X, Xu W, Shi H. Isolation and identification of an inulinase-producing strain and the optimization of its fermentation condition. In: Zhang TC, Nakajima M., eds, Adv Appl Biotechnol Springer-Verlag, Berlin, Heidelberg, 2015, 93–107.
[15] Chi ZM, Zhang T, Cao TS, Liu XY, Cui W, Zhao CH. Biotechnological potential of inulin for bioprocesses. Bioresource Technol 2011, 102, 4295–4303.
[16] Kango N, Jain SC. Production and properties of microbial inulinases: Recent advances. Food Biotechnol 2011, 25, 165–212.
[17] Liu GL, Fu GY, Chi Z, Chi ZM. Enhanced expression of the codon-optimized exo-inulinase gene from the yeast *Meyerozyma guilliermondii* in *Saccharomyces* sp. W0 and bioethanol production from inulin. Appl Microbiol Biotechnol 2014, 98, 9129–38.
[18] Paixao SM, Teixeira PD, Silva TP, Teixeira AV & Alves L. Screening of novel yeast inulinases and further application to bioprocesses. 30, 598–606.
[19] Mazutti MA, Skrowonski A, Boni G, Zabot GL, Silva MF, De-Oliveira D, Di-Luccio M, Maugeri-Filho F, Rodrigues MI, Treichel H. Partial characterization of inulinases obtained by submerged and solid-state fermentation using agroindustrial residues as substrates: A comparative study. Appl Biochem 160, 682–93.
[20] Singh RS, Bhermi HK. Production of extracellular exoinulinase from *Kluyveromyces marxianus* YS-1 using root tubers of *Asparagus officinalis*. Bioresour Tech 2008, 99, 7418–23.
[21] Singh RS, Dhaliwal R, Puri M. Production of inulinase from *Kluyveromyces marxianus* YS-1 using root extract of *Asparagus racemosus*. Process 41, 1703–07.
[22] Singh RS, Saini GK. Production of inulinase from raw Dahlia inulin by *Kluyveromyces marxianus* YS-1. J Sci Indust Res 2013, 72, 603–10.
[23] Mahmoud DAR, Mahdy E-SME, Shousha WGH, Hala W, Refaat HW, Abdel-Fattah AF. Raw garlic as a new substrate for inulinase production in comparison to dry garlic. Aust J Bas Appl Sci 2011, 5, 453–62.
[24] Angel SJ, Kavitha C, Vidyadharani G, Roy P, Dhandapani R. Isolation of inulinase producing bacteria from sugarcane soil. Int J Appl Biol Pharma 2012, 3, 320–26.
[25] Fontana JD, Baron M, Diniz AC, Franco VC. Microbial inulinase secretion using chemically modified inulins. Appl Biochem Biotechnol 1994, 45, 257–68.

[26] Ge XY, Zhang WG. Effects of octadecanoylsucrose derivatives on the production of inulinase by *Apergillus niger* SL-09. World J Microbiol Biotechnol 2005, 21, 1633–38.
[27] Zhang L, Zhao C, Ohta WY, Wang Y. Inhibition of glucose on an exoinulinase from *Kluyveromyces marxianus* expressed in *Pichia pastoris*. Process Biochem 2005, 40, 1541–45.
[28] Singh RS, Sooch BS, Puri M. Optimization of medium and process parameters for the production of inulinase from a newly isolated *Kluyveromyces marxianus* YS-1. Bioresour Technol 2007, 98, 2518–25.
[29] Zherebtsov NA, Shelamova SA, Abramova IN. Biosynthesis of inulinases by *Bacillus* bacteria. Appl Biochem Microbiol 2002, 38, 544–48.
[30] Gavrailov S, Ivanova V. Effects of nitrogen and carbon sources on the production of inulinase from strain *Bacillus* sp. SG113, Acta Scientifica Naturalis 2016, 3, 69–74.
[31] Singh RS, Singh RP, Kennedy JF. Endoinulinase production by a new endoinulinase producer *Aspergillus tritici* BGPUP6 using a low cost substrate. Int J Biol Macromol 2016, 92, 1113–22.
[32] Dinarvand M, Ariff AB, Moeini H, Masomian M, Mousavi SS, Nahavandi R, Mustafa S. Effect of extrinsic and intrinsic parameters on inulinase production by *Aspergillus niger* ATCC 20611. Electro J Biotechnol 2012, 15, 5–5.
[33] Narayanan M, Srinivasan B, Gayathiri A, Ayyadurai A, Mani A. Studies on the optimization and characterization for the biosynthesis of inulinase under solid state fermentation. Int J ChemTech Res 2013, 5, 376–84.
[34] Kango N. Production of inulinase using tap roots of dandelion (*Taraxacum officinale*) by *Aspergillus niger*. J Food Eng 2008, 85, 473–78.
[35] Housseiny MM. Production of an endoinulinase from *Aspergillus niger* AUMC 9375, by solid state fermentation of agricultural wastes, with purification and characterization of the free and immobilized enzyme. J Microbiol 2014, 52, 389–98.
[36] Luo D, Yuan H, Zeng X, Liu J. Optimization of inulinase fermentation conditions of *Aspergillus Niger* X-6 using response surface methodology. In: Bioinformatics and biomedical engineering (iCBBE), 4th International Conference, Chengdu, China, 2010, 1–4.
[37] Abdal-Aziz SAA, El-Metwally MM, Saber WEIA. Molecular identification of a novel inulinolytic fungus isolated from and grown on tubers of *Helianthus tuberosus* and statistical screening of medium components. World J Microbiol Biotechnol 2012, 28, 3245–54.
[38] Saber WIA, El-Naggar NE. Optimization of fermentation conditions for the biosynthesis of inulinase by the new source *Aspergillus tamarii* and hydrolysis of some inulin containing agro-wastes. Biotechnol 2009, 8, 425–33.
[39] El-Aty AAA, Wehaidy HR, Mostafa FA. Optimization of inulinase production from low cost substrates using Plackett–Burman and Taguchi methods. Carbohydr Polym 2014, 102, 261–68.
[40] Singh RS, Singh RP, Yadav M. Molecular and biochemical characterization of a new endoinulinase producing bacterial strain of *Bacillus safensis* AS-08. Biologia 2013, 68, 1028–33.
[41] Gao W, Bao Y, Liu Y, Zhang X, Wang J, An L. Characterization of thermo-stable endoinulinase from a new strain *Bacillus smithii* T7. Appl Biochem Biotechnol 2009, 157, 498–506.
[42] Xiao R, Tanida M, Takao S. Inulinase from *Chrysosporium pannorum*. J Ferment Technol 1988, 66, 553–58.
[43] Sheng J, Chi Z, Li J, Gao L, Gong F. Inulinase production by the marine yeast *Cryptococcus aureus* G7a and inulin hydrolysis by the crude inulinase. Process Biochem 2007, 42, 805–11.
[44] Sheng J, Chi Z, Yan K, Wang X, Gong F, Li J. Use of response surface methodology for optimizing process parameters for high inulinase production by the marine yeast *Cryptococcus aureus* G7a in solid-state fermentation and hydrolysis of inulin. Bioprocess Biosyst Eng 2009, 32, 333–39.

[45] Treichel H, Mazutti MA, Maugeri-Filho F, Rodrigues MI. Technical viability of the production, partial purification and characterisation of inulinase using pretreated agroindustrial residues. Bioprocess Biosyst Eng 2009, 32, 425–33.

[46] Singh RS, Dhaliwal R, Puri M. Production of high fructose syrup from *Asparagus* inulin using immobilized exoinulinase from *Kluyveromyces marxianus* YS-1. J Indus Microbiol Biotechnol 2007, 34, 649–55.

[47] Dilipkumar M, Rajamohan N, Rajasimman M. Inulinase production in a packed bed reactor by solid state fermentation. Carbohydr Polym 2013, 96, 196–99.

[48] Lafuente-Castaneda C, Martinez G, Contreras-Esquivel JC, Rodriguez-Herrera R, Prado-Barragan A, Aguilar CN. Inulinase production by *Penicillium citrinum* ESS in submerged and solid-state cultures. Am J Agri Biologi Sci 2012, 7, 129–34.

[49] Singh RS, Chauhan K. Inulinase production from a new inulinase producer, *Penicillium oxalicum* BGPUP-4. Biocatal Agri Biotechnol 2017, 9, 1–10.

[50] Singh RS, Chauhan K. Sequential statistical optimization of lactose-based medium and process variables for inulinase production from *Penicillium oxalicum* BGPUP-4. 3 Biotech 2018, 8, 38.

[51] Dilipkumar M, Rajasimman M, Rajamohan N. Utilization of copra waste for the solid state fermentative production of inulinase in batch and packed bed reactors. Carbohydr Polym 2014, 102, 662–68.

[52] Barthomeuf C, Regerat F, Pourrat H. Production of inulinase by a new mold of *Penicillium rugulosum*. J Ferment Bioeng 1991, 72, 491–94.

[53] Rawat HK, Ganaie MA, Kango N. Production of inulinase, fructosyltransferase and sucrase from fungi on low-value inulin-rich substrates and their use in generation of fructose and fructo-oligosaccharides. Antonie Van Leeuwenhoek 2015, 107, 799–811.

[54] Park JP, Yun JW. Utilization of chicory roots for microbial endoinulinase production. Lett App Microbiol 2001, 33, 183–87.

[55] Ohta K, Suetsugu N, Nakamura T. Purification and properties of an extracellular inulinase from *Rhizopus* sp. strain TN-96. J Biosci Bioeng 2002, 94, 78–80.

[56] Tasar OC, Erdal S, Algur OF. Utilization of Leek (*Allium ampeloprasum var. porrum*) for inulinase production. Prep Biochem Biotechnol 2015, 45, 596–604.

[57] Onilude AA, Fadaunsi IF, Garuba EO. Inulinase production by *Saccharomyces* sp. in solid state fermentation using wheat bran as substrate. Ann Microbiol 2012, 62, 843–48.

[58] Laowklom N, Chantanaphan R, Pinphanichakarn P. Production, purification and characterization of inulinase from a newly isolated *Streptomyces* sp. CP01. Nat Resour 2012, 3, 137–44.

[59] Dilipkumar M, Rajasimman M, Rajamohan N. Enhanced inulinase production by *Streptomyces* sp. in solid state fermentation through statistical designs. 3 Biotech 2013, 3, 509–15.

[60] Ertan F, Ekinci F, Aktaç T. Production of inulinases from *Penicillium spinulosum, Aspergillus parasiticus* NRRL 2999 and *Trichoderma viride*. Pak J Biol Sci 2003, 6, 1332–35.

[61] El-Souod SMA, Mohamed TM, Ali EM, El-Badry MO, El-Keiy MM. Partial purification of extracellular exo-inulinase from *Ulocladium atrum*. J Genet Eng Biotechnol 2014, 12, 15–20.

[62] Park JP, Bae JT, You DJ, Kim BW, Yun JW. Production of inulooligosaccharides from inulin by a novel endoinulinase from *Xanthomonas* sp. Biotechnol Lett 1999, 21, 1043–46.

[63] Gao L, Chi Z, Sheng J, Wang L, Li J, Gong F. Inulinase-producing marine yeasts: Evaluation of their diversity and inulin hydrolysis by their crude enzymes. Microb Ecol 2007, 54, 722–29.

[64] Pandey A. Solid-state fermentation. Biochem Eng J 2003, 13, 81–84.

[65] Selvakumar P, Pandey A. Solid state fermentation for the synthesis of inulinase from *Staphylococcus* sp. and *Kluyveromyces marxianus*. Process Biochem 1999, 34, 851–55.

[66] Ayyachamy M, Khelawan K, Pillay D, Permaul K, Singh S. Production of inulinase by *Xanthomonas campestris pv phaseoli* using onion (*Alliumcepa*) and garlic (*Allium sativum*) peels in solid state cultivation. Lett Appl Microbiol 2007, 45, 439–44.
[67] Dilipkumar M, Rajasimman M, Rajamohan N. Application of statistical design for the production of inulinase by *Streptomyces* sp. using pressmud. Front Chem Sci Eng 2011, 5, 463–70.
[68] Singh RS, Chauhan K, Singh J, Pandey A, Larroche C. Solid-state fermentation of carrot pomace for the production of inulinase by *Penicillium oxalicum* BGPUP-4. Food Technol Biotechnol 2018, 56, 31–39.
[69] Chen HQ, Chen XM, Li Y, Wang J, Jin ZY, Xu XM, Zhao JW, Chen TX, Xie ZJ. Purification and characterisation of exo- and endo-inulinase from *Aspergillus ficuum* JNSP5-06. Food Chem 2009, 115, 1206–12.
[70] Gao J, Xu YY, Yang HM, Xu H, Xue F, Li S, Feng XH. Gene cloning, expression and characterization of an exo-inulinase from *Paenibacillus polymyxa* ZJ-9. Appl Biochem Biotechnol 2014, 173, 1419–30.
[71] Zhou J, Peng M, Zhang R, Li J, Tang X, Xu B, Ding J, Gao Y, Ren J, Huang Z. Characterization of *Sphingomonas* sp. JB13 exo-inulinase: A novel detergent, salt and protease-tolerant exo-inulinase. Extremophiles 2015, 19, 383–93.
[72] Mohamed SA, Salah HA, Moharam ME, Foda MS, Fahmy AS. Characterization of two thermostable inulinases from *Rhizopus oligosporus* NRRL 2710. J Genet Eng Biotechnol 2015, 13, 65–69.
[73] Xiao R, Tanida M, Takao S. Purification and some properties of endoinulinase from *Chrysosporium pannorum*. J Ferment Bioeng 1989, 67, 244–48.
[74] Xiao R, Tanida M, Takao S. Purification and characteristics of two exoinulinases from *Chrysosporium pannorum*. J Ferment Bioeng 1989, 67, 331–34.
[75] Liu Y, Zhou SH, Cheng YR, Chi Z, Chi ZM, Liu GL. Synergistic effect between the recombinant exo-inulinase and endo-inulinase on inulin hydrolysis. J Mol Catal B: Enzy 2016, 128, 27–38.
[76] Jeza S, Maseko SB, Lin J. Purification and characterization of exo-inulinase from *Paenibacillus* sp. d9 Strain. Protein J 2017, 1–12.
[77] Velikova P, Petrov K, Petrova P. The cell wall anchored β-fructosidases of *Lactobacillus paracasei*: Overproduction, purification and gene expression control. Process Biochem 2017, 52, 53–62.
[78] Arjomand MR, Ahmadian G, Habibi-Rezaei M, Hassanzadeh M, Karkhane AA, Moosavi-Movahedi AA, Amanlou M. The importance of the non-active site and non-periodical structure located histidine residue respect to the structure and function of exo-inulinase. Int J Biol Macromol 2017, 98, 542–49.
[79] Ma JY, Cao HL, Tan HD, Hu XJ, Liu WJ, Du YG, Yin H. Cloning, expression, characterization and mutagenesis of a thermostable exoinulinase from *Kluyveromyces cicerisporus*. Appl Biochem Biotechnol 2016, 178, 144–58.
[80] Arjomand MR, Habibi-Rezaei M, Ahmadian G, Hassanzadeh M, Karkhane AA, Asadifar M, Amanlou M. Deletion of loop fragment adjacent to active site diminishes the stability and activity of exo-inulinase. Int J Biol Macromol 2016, 92, 1234–41.
[81] Yuan B, Hu N, Sun J, Wang SA, Li FL. Purification and characterization of a novel extracellular inulinase from a new yeast species *Candida kutaonensis* sp. nov. KRF1T. Appl Microbiol Biotechnol 2012, 96, 1517–26.
[82] Fawzi EM. Comparative study of two purified inulinases from thermophile *Thielavia terrestris* NRRL 8126 and mesophile *Aspergillus foetidus* NRRL 337 grown on *Cichorium intybus* l. Braz J Microbiol 2011, 42, 633–49.

[83] Kumar VV, Premkumar MP, Sathyaselvabala VK, Dineshkirupha S, Nandagopal J, Sivanesan S. *Aspergillus niger* exo-inulinase purification by three phase partitioning. Eng Life Sci 2011, 11, 607–14.
[84] Coitinho JB, Guimaraes VM, De-Almeida MN, Falkoski DL, De-Queiroz JH, De-Rezende ST. Characterization of an exoinulinase produced by *Aspergillus terreus* CCT 4083 grown on sugar cane bagasse. J Agr Food Chem 2010, 58, 8386–91.
[85] Singh RS, Dhaliwal R, Puri M. Partial purification and characterization of exoinulinase from *Kluyveromyces marxianus* YS-1 for preparation of high-fructose syrup. J Microbiol Biotechnol 2007, 17, 733–38.
[86] Singh RS, Dhaliwal R, Puri M. Development of a stable continuous flow immobilized enzyme reactor for the hydrolysis of inulin. J Indus Microbiol Biotechnol 2008, 35, 777–82.
[87] Singh RS. Enzymatic preparation of high fructose syrup from inulin. In: Panesar PS, Sharma HK, Sarkar BC., eds. Bioprocessing of Foods. Asiatech Publishers Inc, New Delhi, India, 2011, 77–98.
[88] Kim S, Park JM, Kim CH. Ethanol production using whole plant biomass of Jerusalem artichoke by *Kluyveromyces marxianus* CBS1555. Appl Biochem Biotechnol 2013, 169, 1531–45.
[89] Zhao CH, Cui W, Liu XY, Chi ZM, Madzak C. Expression of inulinase gene in the oleaginous yeast *Yarrowia lipolytica* and single cell oil production from inulin-containing materials. Metabol Eng 2010, 12, 510–17.
[90] Cui W, Wang Q, Zhang F, Zhang SC, Chi ZM. Madzak C. Direct conversion of inulin into single cell protein by the engineered *Yarrowia lipolytica* carrying inulinase gene. Process Biochem 2011, 46, 1442–48.
[91] Liu XY, Chi Z, Liu GL, Wang F, Madzak C, Chi ZM. Inulin hydrolysis and citric acid production from inulin using the surface-engineered *Yarrowia lipolytica* displaying inulinase. Metabol Eng 2010, 12, 469–76.
[92] Choi HY, Ryu HK, Park KM, Lee EG, Lee H, Kim SW, Choi ES. Direct lactic acid fermentation of Jerusalem artichoke tuber extract using *Lactobacillus paracasei* without acidic or enzymatic inulin hydrolysis. Bioresource Technol 2012, 114, 745–47.
[93] Wei W, Wu K, Qin Y, Xie Z, Zhu X. Intergeneric protoplast fusion between *Kluyveromyces* and *Saccharomyces cerevisiae* to produce sorbitol from Jerusalem artichokes. Biotechnol Lett 2001, 23, 799–803.
[94] Saha BC. Production of mannitol from inulin by simultaneous enzymatic saccharification and fermentation with *Lactobacillus intermedius* NRRL B-3693. Enzy Microb Technol 2006, 39, 991–95.
[95] Ma ZC, Liu NN, Chi Z, Liu GL, Chi ZM. Genetic modification of the marine-isolated yeast *Aureobasidium melanogenum*. Marine Biotechnol 2015, 17, 511–22.
[96] Li D, Dai JY, Xiu ZL. A novel strategy for integrated utilization of Jerusalem artichoke stalk and tuber for production of 2, 3-butanediol by *Klebsiella pneumoniae*. Bioresour Technol 2010, 101, 8342–47.
[97] Bender H, Wallenfels K. Investigations on pullulan. II. Specific degradation by means of a bacterial enzyme. Biochem Z 1961, 334, 79–95.
[98] Kahar UM, Chan KG, Sani MH, Noh NIM, Goh KM. Effects of single and co-immobilization on the product specificity of type I pullulanase from *Anoxybacillus* sp. SK3-4. Int J Biol Macromol 2017, 104, 322–32.
[99] Lu Z, Hu X, Shen P, Wang Q, Zhou Y, Zhang G, Ma Y. A pH-stable, detergent and chelator resistant type I pullulanase from *Bacillus pseudofirmus* 703 with high catalytic efficiency. Int J Biol Macromol 2018, 1302–1310.

[100] Song W, Nie Y, Mu XQ, Xu Y. Enhancement of extracellular expression of *Bacillus naganoensis* pullulanase from recombinant *Bacillus subtilis*: Effects of promoter and host. Protein Expr Purif 2016, 124, 23–31.

[101] Waleed M, Faiza AA, Nizar I. Production optimization of pullulanase enzyme produced by *Bacillus cereus* isolated from Syrian sources. Int Food Res J 2015, 22, 1824–30.

[102] Duan X, Chen J, Wu J. Improving the thermostability and catalytic efficiency of *Bacillus deramificans* pullulanase by site-directed mutagenesis. Appl Environ Microbiol 2013, 79, 4072–77.

[103] Shankar R, Madihah MS, Shaza EM, Ko NA, Suraini AA, Kamarulzaman K. Application of different feeding strategies in fed batch culture for pullulanase production using sago starch. Carbohydr Poly 2014, 102, 962–69.

[104] Chen WB, Nie Y, Xu Y. Signal peptide-independent secretory expression and characterization of pullulanase from a newly isolated *Klebsiella variicola* SHN-1 in *E. coli*. Appl Biochem Biotech 2013, 169, 41–54.

[105] Liu J, Liu Y, Yan F, Jiang Z, Yang S, Yan Q. Gene cloning, functional expression and characterisation of a novel type I pullulanase from *Paenibacillus barengoltzii* and its application in resistant starch production. Protein Expr Purf 2016, 121, 22–30.

[106] Wu SJ, Chen J. Preparation of maltotriose from fermentation broth by hydrolysis of pullulan using pullulanase. Carbohydr Polym 2014, 107, 94–7.

[107] Singh RS, Saini GK. Pullulan-hyperproducing color variant strain of *Aureobasidium pullulans* FB-1 newly isolated from phylloplane of *Ficus* sp. Bioresour Technol 2008, 99, 3896–99.

[108] Singh RS, Saini GK, Kennedy JF. Pullulan: Microbial sources, production and applications. Carbohydr Polym 2008, 73, 515–31.

[109] Singh RS, Singh H, Saini GK. Response surface optimization of the critical medium components for pullulan production by *Aureobasidium pullulans* FB-1. Appl Biochem Biotechnol 2009, 152, 42–53.

[110] Singh RS, Kaur N, Kennedy JF. Pullulan and pullulan derivatives as promising biomolecules for drug and gene targeting. Carbohydr Polym 2015, 123, 190–207.

[111] Singh RS, Kaur N, Rana V, Kennedy JF. Pullulan: A novel molecule for biomedical applications. Carbohydr Polym 2017, 171, 102–121.

[112] Zou C, Duan X, Wu J. Enhanced extracellular production of recombinant *Bacillus deramificans* pullulanase in *E. coli* through induction mode optimization and a glycine feeding strategy. Bioresour Technol 2014, 172, 174–79.

[113] Duan X, Chen J, Wu J. Optimization of pullulanase production in *Escherichia coli* by regulation of process conditions and supplement with natural osmolytes. Bioresour Technol 2013, 146, 379–85.

[114] Nie Y, Yan W, Xu Y, Chen WB, Mu XQ, Wang X, Xiao R. High-level expression of *Bacillus naganoensis* pullulanase from recombinant *Escherichia coli* with auto-induction: Effect of lac operator. PLoS one 2013, 8, e78416.

[115] Chen WB, Nie Y, Xu Y, Xiao R. Enhancement of extracellular pullulanase production from recombinant *Escherichia coli* by combined strategy involving auto-induction and temperature control. Bioprocess Biosyst Eng 2014, 37, 601–08.

[116] Kang J, Park KM, Choi KH, Park CS, Kim GE, Kim D, Cha J. Molecular cloning and biochemical characterization of a heat-stable type I pullulanase from *Thermotoga neapolitana*. Enz Microb Tech 2011, 48, 260–66.

[117] Chen A, Li Y, Nie J, McNeil B, Jeffrey L, Yang Y, Bai Z. Protein engineering of *Bacillus acidopullulyticus* pullulanase for enhanced thermostability using *in silico* data driven rational design methods. Enz Microb Technol 2015, 78, 74–83.

[118] Li SF, Xu JY, Bao YJ, Zheng HC, Song H. Structure and sequence analysis-based engineering of pullulanase from *Anoxybacillus* sp. LM18-11 for improved thermostability. J Biotechnol 2015, 210, 8–14.
[119] Kunamneni A, Singh S. Improved high thermal stability of pullulanase from a newly isolated thermophilic *Bacillus* sp. AN-7. Enz Microb Technol 2006, 39, 1399–1404.
[120] Nisha M, Satyanarayana T. Characteristics of thermostable amylopullulanase of *Geobacillus thermoleovorans* and its truncated variants. Int J Biol Macromol 2015, 76, 279–91.
[121] Zareian S, Khajeh K, Ranjbar B, Dabirmanesh B, Ghollasi M, Mollania N. Purification and characterization of a novel amylopullulanase that converts pullulan to glucose, maltose, and maltotriose and starch to glucose and maltose. Enzym Microb Technol 2010, 46, 57–63.
[122] Singh RS, Saini GK. Biosynthesis of pullulan and its applications in food and pharmaceutical industry. In: Satyanarayana T, Johri B, Prakash A., eds. Microorganisms in Sustainable Agriculture and Biotechnology. Springer, Dordrecht, 2012, 509–53.
[123] Tonozuka T, Mogi SI, Shimura Y, Ibuka A, Sakai H, Matsuzawa H, Sakano Y, Ohta T. Comparison of primary structures and substrate specificities of two pullulan-hydrolyzing α-amylases, TVA I and TVA II, from *Thermoactinomyces vulgaris* R-47. Biochimica et Biophysica Acta (BBA)-Protein Struct Mol Enzym 1995, 1252, 35–42.
[124] Kim IC, Cha JH, Kim JR, Jang SY, Seo BC, Cheong TK, Lee DS, Choi YD, Park KH. Catalytic properties of the cloned amylase from *Bacillus licheniformis*. J Biol Chem 1992, 267, 22108–114.
[125] Svensson B. Protein engineering in the α-amylase family: Catalytic mechanism, substrate specificity and stability. Plant Mol Biol 1994, 25, 141–57.
[126] Doman-Pytka M, Bardowski J. Pullulan degrading enzymes of bacterial origin. Crit Rev Microbiol 2004, 30, 107–21.
[127] Saha BC, Zeikus JG. Novel highly thermostable pullulanase from thermophiles. Trend Biotechnol 1989, 7, 234–39.
[128] Ramesh B, Reddy PRM, Seenayya G, Reddy G. Effect of various flours on the production of thermostable β-amylase and pullulanase by *Clostridium thermosulfurogenes* SV2. Bioresour Technol 2001, 76, 169–71.
[129] Reddy RM, Reddy PG, Seenayya G. Enhanced production of thermostable β-amylase and pullulanase in the presence of surfactants by *Clostridium thermosulfurogenes* SV2. Process Biochem 1999, 34, 87–92.
[130] Noorwez SM, Ezhilvannan M, Satyanarayana T. Production of a high maltose-forming, hyperthermostable and $Ca^{2+}$ independent amylopullulanase by an extreme thermophile *Geobacillus thermoleovorans* in submerged fermentation. Int J Biotechnol 2006, 5, 337–45.
[131] Kar S, Ray RC, Mohapatra UB. Purification, characterization and application of thermostable amylopullulanase from *Streptomyces erumpens* MTCC 7317 under submerged fermentation. Ann Microbiol 2012, 62, 931–37.
[132] Khalaf AK, Aldeen SB. Optimum condition of pullulanase production by liquid state and solid state fermentation (SSF) method from *Bacillus licheniforms* (BS18). Iraq J Sci 2013, 54, 35–49.
[133] Reddy PRM, Reddy G, Seenayya G. Production of thermostable pullulanase by *Clostridium thermosulfurogenes* SV2 in solid-state fermentation: Optimization of nutrients levels using response surface methodology. Bioprocess Eng 1999, 21, 497–503.
[134] Reddy PRM, Reddy G, Seenayya G. Production of thermostable β-amylase and pullulanase by *Clostridium thermosulfurogenes* SV2 in solid-state fermentation: Screening of nutrients using Plackett-Burman design. Bioprocess Eng 1999, 21, 175–79.
[135] Mrudula S, Reddy G, Seenayya G. Effect of substrate and culture conditions on the production of amylase and pullulanase by thermophilic *Clostridium thermosulforegenes* SVM17 in solid state fermentation. Malays J Microbiol 2011, 7, 19–25.

[136] Nisha M, Satyanarayana T. Characteristics, protein engineering and applications of microbial thermostable pullulanases and pullulan hydrolases. Appl Microbiol Biotechnol 2016, 100, 5661–79.

[137] Kahar UM, Ng CL, Chan KG, Goh KM. Characterization of a type I pullulanase from *Anoxybacillus*. Appl Microbiol Biotechnol 2016, 100, 6291–6307.

[138] Wu H, Yu X, Chen L, Wu G. Cloning, overexpression and characterization of a thermostable pullulanase from *Thermus thermophilus* HB27. Protein Expr Purif 2014, 95, 22–7.

[139] Lee YS, Seo SH, Yoon SH, Kim SY, Hahn BS, Sim JS, Koo BS, Lee CM. Identification of a novel alkaline amylopullulanase from a gut metagenome of *Hermetia illucens*. Int J Biolog Macromol 2016, 82, 514–21.

[140] Orhan N, Kiymaz NA, Peksel A. A novel pullulanase from a fungus *Hypocrea jecorina* QM9414: Production and biochemical characterization. Ind J Biochem Biophy 2014. 51, 149–55.

[141] Xu J, Ren F, Huang CH, Zheng Y, Zhen J, Sun H, Ko TP, He M, Chen CC, Chan HC, Guo RT. Functional and structural studies of pullulanase from *Anoxybacillus* sp. LM18-11. Proteins: Struct Funct Bioinfor 2014, 82, 1685–93.

[142] Wei W, Ma J, Guo S, Wei DZ. A type I pullulanase of *Bacillus cereus* Nws-bc5 screening from stinky tofu brine: Functional expression in *E. coli and Bacillus subtilis* and enzyme characterization. Process Biochem 2014, 49, 1893–1902.

[143] Guan Q, Guo X, Han T, Wei M, Jin M, Zeng F, Liu L, Li Z, Wang Y, Cheong GW, Zhang S. Cloning, purification and biochemical characterization of an organic solvent, detergent, and thermo-stable amylopullulanase from *Thermococcus kodakarensis* KOD1. Process Biochem 2013, 48, 878–84.

[144] Mun SH, Shin M. Mild hydrolysis of resistant starch from maize. Food Chem 2006, 96, 115–21.

[145] Machida Y, Fukui F, Komoto T. Use of oligosaccharides for promoting the proliferation of bifidobacteria. European Patent 1991, 242459.

[146] Goddard P, Nordisk AN. Enzyme at work: A concise guide to industrial enzymes and their uses. Novo Nordisk A/S(2000).

[147] Singh RS, Saini GK, Kennedy JF. Maltotriose syrup preparation from pullulan using pullulanase. Carbohydr Polym 2010, 80, 401–07.

[148] Singh RS, Saini GK, Kennedy JF. Covalent immobilization and thermodynamic characterization of pullulanase for the hydrolysis of pullulan in batch system. Carbohydr Polym 2010, 81, 252–59.

[149] Singh RS, Saini GK, Kennedy JF. Continuous hydrolysis of pullulan using covalently immobilized pullulanase in a packed bed reactor. Carbohydr Polym 2011, 83, 672–75.

[150] Schallmey M, Singh A, Ward OP. Developments in the use of *Bacillus* species for industrial production. Canad J Microbiol 2004, 50, 1–17.

[151] Nisha M, Satyanarayana T. Characterization and multiple applications of a highly thermostable and $Ca^{2+}$-independent amylopullulanase of the extreme thermophile *Geobacillus thermoleovorans*. Appl Biochem Biotechnol 2014, 174, 2594–2615.

Balwinder Singh Sooch, Yogita Lugani and Ram Sarup Singh

# 2 Agro-industrial lignocellulosic residues for the production of industrial enzymes

**Abstract:** Recently, the utilization of agro-industrial lignocellulosic waste materials has received much attention for the production of value-added products like enzymes, organic acids and biofuels. Microbial-based environment-friendly approaches are receiving greater attention for the management of these waste materials in both developing and developed countries due to the increase in problems associated with industrialization and environmental pollution. Agro-industrial lignocellulosic biomass is renewable, inexpensive and abundant and hence can be used as a unique natural source for the cost-effective and large-scale production of these valuable products. Microbial and enzyme technology is one of the most promising approaches in the field of industrial biotechnology for the production of numerous commercial products. Currently, enzymes are extensively used in various industrial sectors like food, pharmaceutics, odontology, textile, detergent, bioremediation, paper and pulp, and wine and brewery. Many industries focused on the utilization of agro-based wastes containing lignocellulosic material as a substrate for the large-scale production of industrial enzymes through either solid-state or submerged fermentation. However, biomass transport, proper handling and pretreatment methods for delignification are some of the limitations associated with the use of lignocellulosic materials. Recently, there has been a resurgence in the utilization of various agricultural wastes like sugarcane baggase, wheat straw, wheat bran, rice straw, rice bran, corn cobs, corn stover, banana waste, hardwood, softwood, newspaper, office paper, groundnut shell, coffee by-products, sweet sorghum and grasses for the cost-effective production of enzymes with better yield and novel activity, which can help to bring sustainability in the agriculture sector. Further improvements in enzyme technology can be achieved by incorporating desirable characteristics through some latest techniques like *in silico* characterization, regulation of metabolic flux, recombinant DNA technology and microarray technology. This chapter describes the biotechnological importance of the utilization of lignocellulosic biomass for the production of industrial enzymes.

**Keywords:** Lignocellulose, industrial enzymes, value addition, biocatalyst, agrowaste.

https://doi.org/10.1515/9783110563337-002

## 2.1 Introduction

Lignocellulosic materials including grasses, woody materials, agricultural wastes and forestry residues mainly comprise lignin (10–25%), cellulose (40–50%) and hemicellulose (20–30%) [1, 2]. They act as promising feedstock for the smooth functioning of modern biotechnological industrial sectors including food and feed, chemicals, pulp and paper, plastics, bioprocessing catalysts, textiles, and biofuels. [3]. A large amount of lignocellulosic biomass is produced annually in the form of agriculture waste as well as from agro-based industries [4]. The increase in global population and improved economies in many countries have resulted in enhanced global consumption, thereby adversely affecting the environment [5, 6]. Therefore, to fulfill human needs without adversely affecting natural resources, there is an utmost need to reduce the impact per unit of product produced [7]. Most of the lignocellulosic biomass in developing and developed countries are disposed of by burning, which has adverse effects on the environment. Therefore, there is an urgent need to utilize this cheap lignocellulosic biomass for the production of value-added products like biofuels, organic acids and enzymes. This approach will not only create a safe environment but also contribute to sustainable agriculture [8–11].

Among the various biotechnological industrial sectors, enzymatic processing is one of the most promising approaches, and it has shown a wide range of industrial applications. Enzymes act as biological catalysts for many biochemical reactions, and their unique properties like substrate specificity, stereospecificity, pH optima and temperature optima make them good candidates for their use in many industrial products. In the current few decades, many industries have been focusing on the isolation of extremophilic microorganisms, which remain stable under adverse environmental conditions like high temperature and alkaline and halophilic conditions. Industries are also developing some cheaper enzyme production methods with enhanced yield and productivity to reduce the cost of enzymes [12]. Microorganisms like bacteria, yeast and fungi are grown under submerged fermentation or solid-state fermentation (SSF) conditions for the production of enzymes, which may be either extracellular or intracellular. Furthermore, diverse downstream processing approaches have been utilized for the extraction and purification of intracellular enzymes. The high cost of substrates for media acts as a major constraint in the microbial production of industrial enzymes. However, the abundant availability of these lignocellulosic agrowaste makes them good alternative as substrates for the production of industrial enzymes with or without pretreatments. Hence, this chapter aims to emphasize the scope and biotechnological importance of lignocellulosic biomass for the production of different industrial enzymes along with their market scenario.

## 2.2 Biotechnological importance of lignocellulosic biomass

Lignocellulosic biomass, comprising cellulose (major structural component of the plant cell wall providing mechanical strength), hemicellulose (repeated polymers of pentoses and hexoses) and lignin (containing aromatic alcohols), provide a protective seal around cellulose and hemicellulose [13, 14]. The structure of lignocellulosic material containing cellulose, hemicellulose and lignin is shown in Figure 2.1, and the composition of lignocellulosic components in various lignocellulosic materials is listed in Table 2.1.

Lignocellulosic biomass acts as potential candidates for the biotechnological production of many value-added products like biofuels, enzymes and organic acids. [15]. With increased industrialization, the recent years have seen the development of novel technologies for the recycling of agrowaste containing lignocellulosic material through green biotechnology for varied purposes. However, large amount of agriculture waste produced by many countries is still not properly utilized; instead it is destroyed through burning, thus causing environmental pollution [16]. The conversion of this lignocellulosic material into high-value-added products is a multistep process including pretreatment (mechanical, chemical or biological), enzymatic hydrolysis and fermentation process [17]. A schematic representation of the utilization of lignocellulosic material for microbial production of industrial enzymes along with their applications has been provided in Figure 2.2.

Table 2.1: Composition of lignocellulosic components in agro-industrial lignocellulosic waste [1, 2].

| Lignocellulosic material | Cellulose (%, w/w) | Hemicellulose (%, w/w) | Lignin (%, w/w) |
|---|---|---|---|
| Wheat straw | 29–35 | 26–32 | 16–21 |
| Rice straw | 38.0 | 26.0 | 19.0 |
| Sugarcane baggase | 42.0 | 25.0 | 20.0 |
| Corn cobs | 45.0 | 35.0 | 15.0 |
| Corn stover | 38.0 | 26.0 | 19.0 |
| Newspaper | 40–55 | 25–40 | 18–30 |
| Hardwood | 40–55 | 24–40 | 18–25 |
| Softwood | 45–50 | 25–35 | 25–35 |
| Banana waste | 13.2 | 14.8 | 14.0 |
| Grasses | 25–40 | 25–50 | 10–30 |

## 2.3 Utilization of lignocellulosic materials for the production of industrial enzymes

Numerous studies have focused on the utilization of various agricultural lignocellulosic waste materials, such as sugarcane baggase, wheat straw, wheat bran, rice

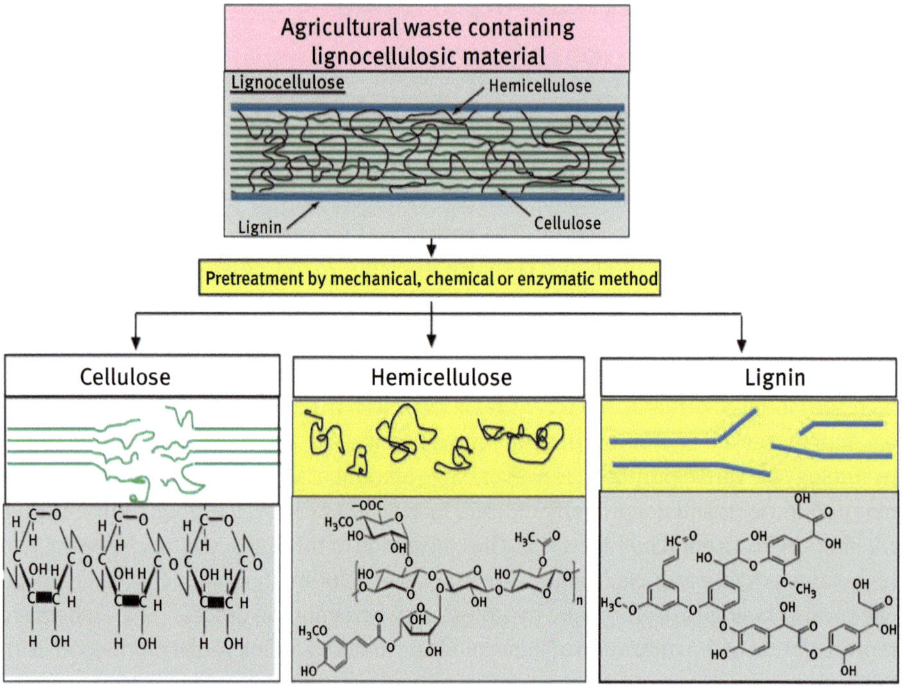


Figure 2.1: Structure of lignocellulosic material.

straw, rice bran, corn cobs, corn stover, banana waste, hardwood, softwood, newspaper, office paper, groundnut shell, coffee by-products, sweet sorghum and grasses, for the production of industrially important enzymes like protease, cellulase, lipase, xylanase, pectinase, amylase, glucose oxidase, inulinase, lipooxygenase, phytase, peroxidase and xylose reductase. A list of the utilization of various lignocellulosic agricultural wastes for the production of industrial enzymes is provided in Table 2.2.

Those agricultural wastes, which act as energy-rich substrates for microbial fermentation, are preferred at the industrial level due to their low cost and lesser problems associated with their waste disposal. They are also considered the best substrate for microorganisms during the process of SSF for the production of industrial enzymes. Several factors influence the selection of agricultural residues for enzyme production, such as nutrient composition, cost and availability [37]. After selecting a suitable substrate, the next important parameter is its pretreatment using various mechanical, chemical and enzymatic methods to make it easily accessible for microbial growth. Various nutritional (media composition) and process (temperature, pH, aeration, agitation, etc.) parameters affect the fermentation process during the production of enzymes, and these need to be tailored to increase the efficiency of the bioprocess [38]. Although these process

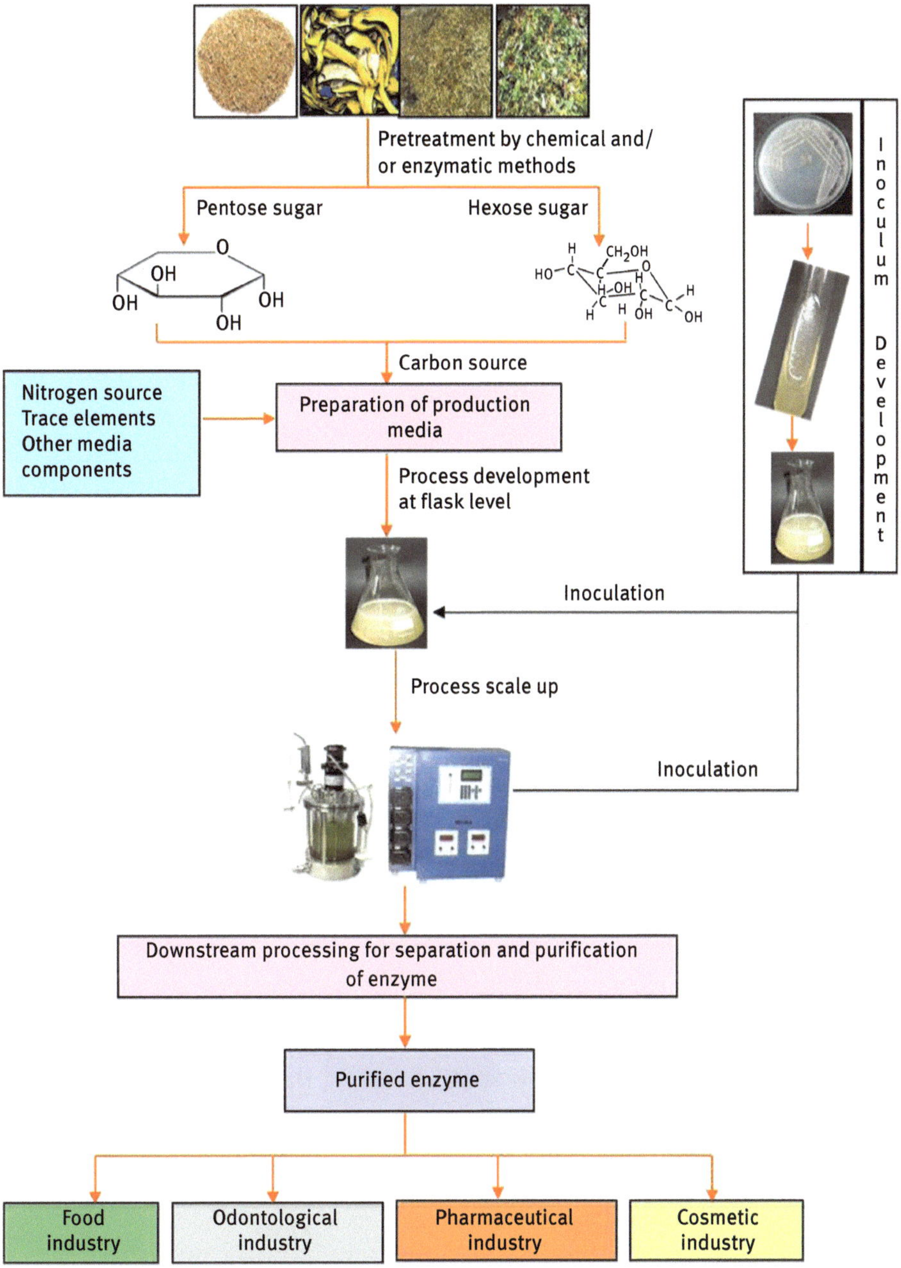


Figure 2.2: Schematic representation of the utilization of lignocellulosic material for the microbial production of industrial enzymes along with their applications.

Table 2.2: List of agro-industrial lignocellulosic materials for the production of different industrial enzymes.

| Lignocellulosic material | Microbial culture | Enzymes produced | Reference |
|---|---|---|---|
| Rice bran | *Bacillus sp.* JB-99 | Xylanase | [18] |
| Rice bran | *Bacillus sp.* | Protease | [19] |
| Coconut oil cake and wheat bran | *Aspergillus sp.* | Protease | [20] |
| Agava-tequilana waste | *Aspergillus niger* | Xylanase | [21] |
| Coffee by-products | *Aspergillus oryzae* | Protease | [22] |
| Groundnut oil cakes | *Bacillus subtilis* | Lipase | [23] |
| Oil palm empty fruit bunch | *Aspergillus terreus* | Cellulase | [24, 25] |
| Chicory Dahlia | *Penicillium citrinum* | Inulinase | [26] |
| Wheat bran | *Aspergillus niger* | Phytase | [27] |
| Empty fruit bunches, rice straw, rice husk, corn cob | *Marasmius sp., Trametes hirsute, Trametes versicolor, Phanerochaete crysosporium* | Laccase | [28] |
| Wheat bran | *Bacillus stearothermophilus* | Lipase | [29 |
| Pineapple<br>Banana waste | *Aspergillus flavus*<br>Bacterial consortia (*Bacillus megaterium, Cellulomonas cartae, Pseudomonas putida, Pseudomonas fluorescence*) | Pectinase<br>Cellulase | [30]<br>[31] |
| Sugarcane baggase, sisal fiber, coconut shell | *Trametes villosa* | Manganese peroxidase | [32] |
| Wheat straw, paddy straw, maize straw, sugarcane baggase | *Bacillus licheniformis* MTCC 1483 | Amylase | [33] |
| Corn cob | *Candida tropicalis* | Xylose reductase | [34] |
| Orange peel | *Bacillus licheniformis* | Pectinase | [35] |
| Carrot pomace | *Penicillium oxalicum* | Inulinase | [36] |

parameters can be easily regulated during the submerged fermentation process, but it is difficult to regulate them during SSF. Therefore, for the commercial production of enzymes, it is very important to optimize all the parameters affecting the production of enzymes [39, 40].

Extracellular enzymes are directly released during the fermentation process, whereas for intracellular enzymes, extra steps are required for extraction from the microbial cells. Various cell-disruption methods like physical processes (freeze-thawing, mortar and pestle, sonication, glass beads), chemical processes (using ethylenediaminetetraacetic acid, β-mercaptoethanol, etc.) and enzymatic processes (using lysozymes and nanoparticles) are commonly employed for the disruption of microbial cells. Enzyme-purification techniques like dialysis, ultrafiltration, gel filtration and affinity chromatography further increase the cost of enzymes [41]. Therefore, easy and cost-effective techniques need to be developed for the production and purification of enzymes for commercializing these biocatalysts at the industrial scale. Although many patents have already been granted on the production of various enzymes, there is still a need to focus on economically viable production of these industrially important enzymes [2, 42, 43]. The microbial sources and industrial applications of enzymes have been listed in Table 2.3.

Research on the isolation of new and efficient extremophilic microorganisms like psychrophiles, alkalophiles, acidophiles and halophiles from environmental and industrial niches is needed for enzyme production because of their advantages like their resistance toward environmental stress and their stability under harsh conditions [46, 47]. Some progress has already been made in the field of enzyme

Table 2.3: Microbial sources and industrial applications of enzymes [44, 45].

| Enzyme | Microbial source | Industrial Application |
|---|---|---|
| Cellulase | *Bacillus subtilis*<br>*Clostridium acetobutylicum*<br>*Streptomyces lividans*<br>*Aspergillus niger*<br>*Fusarium solani* | Textile<br>Laundry and detergent<br>Food and animal feed<br>Paper and pulp industry<br>Biofuel |
| Xylanase | *Clostridium stercorarium*<br>*Neurosposra crassa*<br>*Thermomonospora curvata*<br>*Aspergillus awamori*<br>*Trichoderma hamatum* | Baking<br>Paper and pulp<br>Animal feed |

Table 2.3 (continued)

| Enzyme | Microbial source | Industrial Application |
|---|---|---|
| Pectinase | *Aspergillus niger*<br>*Rhizopus stolonifer*<br>*Fusarium oxysporum*<br>*Penicillium citrinum* | Textile<br>Beverage |
| Protease | *Bacillus subtilis*<br>*Serratia marcescens*<br>*Aspergillus flavus*<br>*Saccharmyces lipolytica*<br>*Yarrowia lipolytica* | Food<br>Detergent<br>Leather<br>Medicine<br>Brewing |
| Lipase | *Bacillus subtilis*<br>*Pseudomonas aeruginosa*<br>*Streptomyces grisesus*<br>*Mucor sp.* | Baking<br>Detergent<br>Dairy<br>Textile |
| Amylase | *Bacillus coagulans*<br>*Bacillus subtilis*<br>*Cryptococcus flavus*<br>*Aspergillus oryzae* | Detergent<br>Baking<br>Paper and pulp<br>Textile |
| Glucose oxidase | *Aspergillus flavus*<br>*Penicillium sp.*<br>*Fusarium sp.* | Baking |
| Phytase | *Bacillus sp.*<br>*Pseudomonas sp.*<br>*Aspergillus niger* | Animal feed |
| Peroxidase | *Bacillus subtilis*<br>*Pseudomonas sp.*<br>*Candida krusei*<br>*Phanerochaete chrysosporium* | Textile |
| Xylose reductase | *Enterobacter sp.*<br>*Candida boidinii*<br>*Aspergillus sp.*<br>*Neurospora crassa* | Xylitol production |

technology in the past few decades. Various genome and protein computational tools are used for the characterization of industrially important enzymes using *in silico* studies to obtain the desirable products. Some research is also under progress on the production of genetically modified microorganisms using recombination

DNA technology with enhanced enzyme-producing ability using microarray technology by understanding genomics, proteomics, metabolomics and system biology. The status of the microbial production of some of the industrially important enzymes like cellulase, xylanase, inulinase, xylose reductase, protease and lipase from lignocellulosic materials has been discussed below.

### 2.3.1 Cellulases

Cellulases hydrolyze cellulose to produce glucose, cellobiose and cellooligosaccharides. There are three major types of cellulases, namely, cellobiohydrolase (EC 3.2.1.91), endo-β-1,4-glucanase (EC 3.2.1.14) and β-glucosidase (EC 3.2.1.21). Cellulases have been produced by many microorganisms including bacteria (*Bacillus, Clostridium, Pseudomonas, Acidothermus*), fungi (*Aspergillus, Fusarium, Trichoderma, Penicillium, Humicola*) and actinomycetes (*Cellulomonas, Streptomyces, Thermononospora*); however, commercial production of this enzyme is mainly carried out from fungal cultures using lignocellulosic waste under SSF [24, 25, 44, 48].

Various studies have been reported for the production of cellulases utilizing lignocellulosic hydrolysates such as cassava waste [49], paddy straw [50], palm oil mill waste [51], oil palm decanter cake [52], rice husks, millet husks, wheat leaves, wheat straws, maize cobs, maize strips [53], corn cob, cane baggase, sawdust [54], banana wastes [31], wheat bran, rice bran, maize bran [55], groundnut [56], rice straw [57], soybean hulls, wastepaper [58] and fruit peels [59]. They have shown a wide range of enormous potential applications in different industries like food, pharmaceutical, brewery and wine, textile, detergent, and paper and pulp [60–62].

### 2.3.2 Xylanases

Xylanases are glycosidases (EC 3.2.1.x), and they are highly diverse with different specificities, primary sequences and folds [63]. These enzymes catalyze the hydrolysis of xylan at 1,4-β-D-xylosidic linkages. Xylanases have been produced by many bacteria (*Bacillus, Cellulomonas, Acidobacterium, Staphylococcus*), molds (*Aspergillus, Penicillium, Trichoderma, Fusarium, Cephalosporium*), yeasts (*Cryptococcus*) and actinomycetes (*Streptomyces, Thermomyces, Thermomonospora*).

Different lignocellulosic materials, like sawdust of Abora wood, dry maize [64], sugarcane baggase, rice husk, soybean meal, corn cobs, wheat straw [65], wheat bran, sawdust [66], apple pomace, sawdust, coffee-processing plant waste, cassava waste, tea production wastes [67], orange waste [68], cassava peels [69], and corn cob [70], have been utilized previously for the production of xylanase. This enzyme is mainly used in wastepaper recycling, food and feed, and textile and pulp industries [71, 72].

## 2.3.3 Inulinases

Inulinolytic enzymes are an important class of industrial enzymes belonging to the glycoside hydrolase (GH) families 32 and 92. Family GH 32 mainly consists of exoinulinase (EC 3.2.1.7), 1-exohydrolase (EC 3.2.1.153), 1,2-B fructan $1^F$-fructosyltransferase (EC 2.4.1.100) and sucrose $1^F$-fructosyltransferase (EC 2.4.1.99). Among these inulinolytic enzymes, exoinulinase and endoinulinase are very important food enzymes that act on inulin to produce fructose and fructooligosaccharides, respectively. Exoinulinases are produced by diverse groups of microorganisms including fungi (*Aspergillus* sp., *Penicillium* sp., *Rhizopus* sp., *Fusarium* sp., *Chrysosporium* sp., etc.), yeasts (*Kluyveromyces* sp., *Candida* sp., *Pichia* sp., *Saccharomyces* sp., etc.), bacteria (*Bacillus* sp., *Lactobacillus* sp., *Pseudomonas* sp., etc.) and actinomycetes (*Streptomyces* sp.). Microbial sources of endoinulinases include *Aspergillus* sp., *Penicillium* sp., *Trichoderma* sp., *Bacillus* sp., *Xanthomonas* sp., *Candida* sp., *Yarrowia* sp. [73]. Inulinases are produced from a wide variety of substrates employing submerged fermentation and SSF. Inulinases are inducible enzymes, and inulin is always used as an inducer for the production of inulinases. Submerged fermentation is the most efficacious and accepted method used for the production of inulinases. However, there are numerous reports on the production of inulinases in SSF using lignocellulosic materials like sugarcane bagasse [74, 75], copra waste [76, 77], dandelion roots and chicory tubers [78], wheat bran and rice husk [79, 80], sunflower tubers and lettuce roots [81], chicory roots, artichoke leaves and banana leaves [82] and carrot pomace [36]. For successful SSF, parameters like substrate particle size, moisture, pH, relative humidity, substrate pretreatment, incubation temperature, agitation, aeration, and size of inoculum are important factors. In addition, heat and mass transfer also play a significant role during SSF. The preparations of high fructose syrup [83–86] and fructooligosaccharides [62, 87] are two major applications of inulinases in the food industry. A number of reports are available on the preparation of high fructose syrup and fructooligosaccharides at the laboratory scale using either whole cells or soluble/immobilized enzymes in batch and continuous systems [73]. Simultaneous saccharification and fermentation of inulin/inulin-rich plant extracts using different microorganisms has also been reported for the production of various fermentation products like 2-3 butanediol, ethanol, gluconic acid, lactic acid, mannitol, sorbitol, single-cell proteins and pullulan [86].

## 2.3.4 Xylose reductase

Xylose reductase (EC 1.1.1.21) belongs to the oxidoreductase class of enzymes and catalyzes the conversion of xylose into xylitol. This enzyme is mainly present in yeasts (*Candida, Saccharomyces, Debaryomyces, Pichia, Hansenula, Torulopsis, Kloeckera, Trichosporon, Cryptococcus, Rhodotorula, Monilia, Kluyveromyces*) and molds (*Neurospora, Penicillium, Aspergillus, Fusarium, Mucor, Rhizopus*), but a few members of facultative bacteria (*Corynebacterium, Enterobacter*) are also reported for the production of xylose reductase [88]. The industrial importance of this enzyme is due to its ability to produce xylitol, a pentitol sugar [89].

Various studies have been reported on the production of xylose reductase by utilizing lignocellulosic hydrolysates, such as grape pomace, sugar beet pomace, rice and barley straw, cotton waste, forest residues, sunflower stalks, brewer's spent grain [90], sugarcane baggase [91], corn cob [92, 93], banana peel, sweet lime peels, wheat flour, sesame oil cake, groundnut oil cake, ragi flour and pomegranate peels [92]. Xylitol has shown tremendous application in many sectors like food, odontology, pharmaceutics, textiles and cosmetics [94–97].

### 2.3.5 Proteases

Proteases belong to the class hydrolases and are classified in subgroup 4 of group 3. There are many types of proteases like exopeptidases, endopeptidases, aminopeptidases, carboxypeptidases, serine proteases, aspartate proteases and cysteine proteases [98]. This enzyme is highly ubiquitous with a wide range of sources like plants, animals and microorganisms. Many microorganisms belonging to bacteria (*Bacillus, Streptomyces, Serratia*), yeasts (*Saccharomyces, Candida, Yarrowia*) and molds (*Aspergillus*) are used for the production of this enzyme [43].

The production of proteases from different agro-industrial wastes, like wheat bran, rice bran, casein [19], sugarcane baggase, agava tequilana waste, shrimp shell waste, wheat straw, wheat bran, rice bran [43], banana peel, black gram husk, cow dung paddy straw [99], and manure kitchen waste [100], has been reported. Proteases have many applications in the food and detergent industry [101, 102].

### 2.3.6 Pectinases

Pectinases, an industrially growing, heterogeneous group of related enzymes in the biotechnological sector, have shown many promising applications in different industrial sectors such as food and feed, clearing of fruit juices, textile, oil extraction, paper and pulp, and tea and coffee fermentation [103, 104]. Based on the mode of action, pectinases are classified into polygalactouronase (EC 3.2.1.15), pectin esterase (EC 3.1.1.11), pectin lyase (EC 4.2.2.10) and pectin lyase (EC 4.2.2.2). The major sources of pectinases are yeast and fungi (50%), bacteria (35%) and plant/animal (15%) [105, 106].

Pectinases can be produced from different agro-industrial wastes like molokhia stalk [107], wheat bran, sugar beet pulp, sunflower plate, orange peel, banana peel, apple pomace, grape pomace [108], pineapple peel, sawdust, cassava waste [30], sugarcane baggase [109], orange and lemon peels [110]. Pectinases have many applications in fruit processing, textile industry, paper industry, oil extraction and wastewater treatment [104, 111, 112].

### 2.3.7 Lipases

Lipases (EC 3.1.1.3) catalyze the hydrolysis of long-chain triglycerides to release fatty acids and glycerol [113]. They are mainly produced by bacteria (*Pseudomonas, Acinetobacter, Bacillus, Rhizopus, Serratia*), yeasts (*Candida, Rhodotorula, Yarrowia, Saccharomyces*) and molds (*Aspergillus, Rhizopus, Penicillium*).

Various studies have reported the production of lipase from different agro-industrial wastes like mustard oil cake, linseed oil cake [23], canola seed oil cake, peanut shells, sunflower hulls [114], wheat bran, rice bran, groundnut cake, neem cake, coconut cake [115], sugarcane baggase, cassava peel, barley spent grain and citrus pulp [116]. Lipases are mainly used in the fat and oleo-chemical industry, the detergent industry, in food processing, for flavor development and for the production of biodegradable polymers [117–121].

## 2.4 Global and Indian markets of enzymes

The current industrial market of enzymes is growing at a very fast rate, and it is projected to reach US $6.30 B by 2022 [122]. Currently, the demand for industrial enzymes has been found to be stable in developed countries like the United States, Western Europe, Japan and Canada; however, the industrial enzyme market has been growing at a fast rate in many countries of the Asia-Pacific, Eastern Europe, Africa and the Middle East regions. An increase in demand of these enzymes is expected in the pharmaceutical sector in the developing countries for improving the medical care facilities. Bioethanol production is another sector that has been receiving attention by investors for launching fuels with novel and unique properties in the market.

Some enzyme-manufacturing industries in the world are Hayashibara Company, Maps Enzyme Ltd., Nexgen Biotechnologies Inc., Asahi Kasei Pharma, Advanced Enzyme Technologies Ltd., Novozymes A/S, Amano Enzyme Inc., AB Enzymes GmbH, Cargill Texturizing Solutions and Genencor International Inc. The global market status of some industrial important enzymes is listed in Table 2.4.

From 2000 to date, there has been a tremendous increase in the number of patents for industrial enzymes due to the growth of green technology [42]. Of these, the maximum numbers of patents have been for proteases, followed by amylases and cellulases due to their wide use in various industrial sectors. With the development of green technology, the patent status for other enzymes is expected to increase in the

Table 2.4: Global market of industrially important enzymes.

| Enzyme | Application | Company | Potential market value | Reference |
|---|---|---|---|---|
| Protease | Meat | Novozymes | $1.37 M in 2012 | [122] |
| Amylase | Textiles Baking | Novozymes | $2.7 M by 2019 $695.1 M by 2019 | [122] |
| Xylanase | Pulp and paper | Novozymes | 200M | [123] |
| Cellulase, Lipase | Detergent | Novozymes | $2.767 M by 2019 | [122] |
| β-glucanase | Brewing | Carlsberg | $1.37 M by 2020 | [122] |
| Urease | Wine | Carlsberg | $1.37 M by 2020 | [122] |

future. The robustness of many enzymes like kinases, polymerases and nucleases is due to their extreme stability, specificity and stability over a wide range of pH, making them important candidates for use in medicine and biotechnology [124].

The maximum demand for enzymes has been in the pharmaceutical sector (50%), followed by the detergent sector (20%) and the textile industry (20%) and the rest with the food industry [125]. Thus, the operational performance of enzymes for industrial applications should be enhanced using advanced techniques like use of nanoparticles.

## 2.5 Conclusions

Green biotechnology involves the utilization of agro-industrial lignocellulosic materials for the production of industrially important products, mainly enzymes, organic acids and biofuels, to make them efficient and cost effective. This area of research is gaining tremendous impetus by different industrial sectors due to the abundance and easy availability of low-cost agricultural waste. Agricultural wastes usually require some pretreatment to make them utilizable for microbial growth. Furthermore, enzyme technology is one of the fastest growing sectors in the world, and its global market is growing at a very high rate. Many enzymes like amylases, proteases and cellulases have already been utilized in various industrial sectors like food, bakery, detergent, pharmaceutical, textile and biofuels; however, some others like xylose reductase have not been utilized commercially to date. Therefore, there is a need to focus on the large-scale, low-cost production of such novel industrially important enzymes by utilizing various agro-based lignocellulosic waste materials. This will also contribute toward the development of sustainable agriculture by the usage of waste materials in a profitable manner. Some advanced techniques like *in silico* characterization, regulation of metabolic flux, recombinant DNA technology and microarray technology should also be adapted to tailor the properties of these enzymes or microorganisms according to industrial demand.

**Acknowledgements:** Authors are thankful to University Grant Commission, New Delhi, India for providing support in the form of MRP and Bhai Kahn Singh Nabha Library of Punjabi University, Patiala, India for providing access to scientific literature.

## References


[1] Iqbal HMN, Kyazze G, Keshavarz T. Advances in valorization of lignocellulosic materials by biotechnology: An overview. BioResource 2013, 8(2), 3157–3176.

[2] Anwar Z, Gulfraz M, Irshad M. Agro-industrial lignocellulosic biomass a key to unlock the future bio-energy: A brief review. J Radiat Res Appl Sci 2014, 7(2), 163–173.

[3] Saha BC. Hemicellulose bioconversion. J Ind Microbiol Biotechnol 2003, 30(5), 279–291.

[4] Perez J, Rubia TMD, Martinez J. Biodegradation and biological treatments of cellulose, hemicellulose and lignin: An overview. Int Microbiol 2002, 5(2), 53–63.

[5] UNFPA. Population and climate change- Framework of UNFPA's Agenda. United Nations Population Fund Report 2008.

[6] Fischer-Kowalski M, Swilling M, von Weiz-sacker EU, Ren Y, Moriguchi Y, Crane W, Krausmann F, Eisenmenger N, Gilijum S, Hennicke P, Romero-Lankao P, Siriban Manalang, A, Sewerin S. Decoupling natural resource use and environmental impacts from economic growth. United Nations Environment Programme Report of the working group on decoupling of the international resource panel. 2011a.

[7] UNEP. Towards a like style sustainability assessment: making informed choices on products. UNEP/SETAC Life cycle initiative. United Nations Environment Programme Report 2011b, DTI/1412/PA.

[8] Mahmoodi NM, Arami M, Mazaheri F, Rahimi S. Degradation of sericin (degumming) of Persian silk by ultrasound and enzymes as a cleaner and environmentally friendly process. J Clean Prod 2010, 18(2), 146–151.

[9] Isroi MR, Syamsiah S, Niklasson C, Cahyanto MN, Lundquist K, Taherzadeh MJ. Biological pretreatment of lignocelluloses with white rot fungi and its applications: A review. BioResource 2011, 6(4), 5224–5259.

[10] Asgher M, Ahmad Z, Iqbal, HMN. Alkali and enzymatic delignification of sugarcane baggase to expose cellulose polymers for saccharification and bioethanol production. Ind Crop Prod 2013, 44, 488–495.

[11] Ofori-Boateng C, Lee KT. Sustainable utilization of oil palm wastes for bioactive phytochemicals for the benefit of the oil palm and nutraceutical industries. Phytochemistry Rev 2013, 12(1), 173–190.

[12] Mukherjee S, Das P, Sivapathasekaran C, Sen R. Enhanced production of biosurfactant by a marine bacterium on statistical screening of nutritional parameters. Biochem Eng J 2008, 42(3), 254–260.

[13] Jiang G, Nowakowski DJ, Bridgwater AV. A systematic study of the kinetics of lignin pyrolysis. Thermochimica Acta 2010, 498(1), 61–66.
[14] Menon V, Rao M. Trends in bioconversion of lignocelluloses: biofuels, platform chemicals and biorefinery concept. Prog Energy Combust Sci 2012, 38(4), 522–550.
[15] Koutinas AA, Wang RH, Webb C. The biochemurgist-bioconversion of agricultural raw materials for chemical production. Biofuels Bioprod Biorefin 2007, 1(1), 24–38.
[16] Okonko IO, Adeola T, Enobong F, Adewale OA. Utilization of food wastes for sustainable development. Electron J Environ Agric Food Chem 2009, 8(4), 263–286.
[17] Wyman CE, Dale BE, Elander RT, Holtzapple M, Ladlisch MR, Lee YY. Coordinated development of leading biomass pretreatment technologies. Bioresource Technol 2005, 96(18), 1959–1966.
[18] Virupakshi S, Babu OK, Gaikwad SR, Naik GR. Production of a xylanolytic enzyme by a thermo alkalophilic *Bacillus sp.* JB-99 in solid state fermentation. Proc Biochem 2005, 40(1),431–435.
[19] Naidu KSB, Devi KL. Optimization of thermostable alkaline protease production from species of *Bacillus* using rice bran. Afr J Biotechnol 2005, 4(7), 724–726.
[20] Sumantha A, Chandran S, George S, Soccol CR, Pandey A. Production and partial purification of a neutral metalloprotease by fungal mixed substrate fermentation. Food Technol Biotechnol 2005, 43(4), 313–319.
[21] Huitron C, Perez R, Sanchez AE, Lappe P, Zavaleta LR. Agricultural waste from the tequila industry as substrate for the production of commercially important enzymes. J Environ Biol 2008, 29(1), 37–41.
[22] Murthy PS, Naidu MM. Protease production by *Aspergillus oryzae* in solid state fermentation utilizing coffee by products. World Appl Sci J 2010, 8(2), 199–205.
[23] Singh M, Saurav K, Srivastava N, Kannabiran K. Lipase production by *Bacillus subtilis* OCR-4 in solid state fermentation using ground nut oil cakes as substrate. Curr Res J Biol Sci 2010, 2(4), 241–245.
[24] Shahriarinour M, Ramanan RN, Wahab MNA, Mohamad R, Mustafa S, Ariff AB. Improved cellulase production by *Aspergillus terreus* using oil palm empty fruit bunch fibre as substrate in a stirred tank bioreactor through optimization of the fermentation conditions. BioResource 2011, 6(3), 2663–2675.
[25] Shahriarinour M, Wahab MNA, Mustafa S, Mohamad R, Ariff AB. Effect of various pretreatments of oil palm empty fruit bunch fibres for subsequent use as substrate on the performance of cellulase production by *Aspergillus terreus*. BioResource 2011, 6(1), 291–307.
[26] El-Hersh MS, Saber WIA, El-Naggar NEA. Production strategy of inulinase by *Penicillium citrinum* AR-IN2 on some agricultural by products. Microbiol J 2011, 1(3), 79–88.
[27] Kumar M, Sushma. Optimization of production media of phytase from *Aspergillus niger* using wheat bran waste. Int J Sci Res 2012, 3(6), 949–951.
[28] Risdianto H, Sofianti E, Harjati SS, Setiadi T. Optimization of laccase production using white rot fungi and agriculture wastes in solid state fermentation. J Eng Sci 2012, 44(2), 93–105.
[29] Sabat S, Murthy VK, Pavithra M, Mayur P, Chandavar A. Study of enhanced lipase production using agrowaste product by *Bacillus stearothermophilus* MTCC 37. Int J Pharm Chem Biol Sci 2012, 2(3), 266–274.
[30] Thangaratham T, Manimegalai G. Optimization and production of pectinase using agro waste by solid state and submerged fermentation. Int J Curr Microbiol Appl Sci 2014, 3(9), 357–365.
[31] Dabhi BK, Vyas RV, Shelat HN. Use of banana waste for the production of cellulolytic enzymes under solid substrate fermentation using bacterial consortium. Int J Curr Microbiol Appl Sci 2014, 3(1), 337–346.
[32] Silva MLC, Souza VB, Santos VS, Kamids HM, Neto JRTV, Neto AG, Koblitz MGB. Production of manganese peroxide by *Trametes villosa* on unexpensive substrate and its application in the removal of lignin from agricultural wastes. Adv Biosci Biotechnol 2014, 5(14), 1067–1077.

[33] Kaur PS, Kaur S, Kaur H, Sharma A, Raj P, Panwar S. Solid substrate fermentation using agro industrial waste: New approach for amylase production by *Bacillus licheniformis*. Int J Curr Microbiol Appl Sci 2015, 4(12), 712–717.
[34] Barathikannan K, Khusro A, Paul A. Optimization of parameters to increase the xylose reductase production from *Candida tropicalis* strain LY15 using corn cob as hemicellulose waste substrates. Afr J Microbiol Res 2016, 10(45), 1908–1917.
[35] Bibi N, Ali S, Tabassum R. Statistical optimization of pectinase biosynthesis from orange peel by *Bacillus licheniformis* using submerged fermentation. Waste Biomass Volariz 2016, 7(3), 467–481.
[36] Singh RS, Chauhan K. Singh J, Pandey A, Larroche C. Solid state fermentation of carrot pomace for the production of inulinase by *Penicillium oxalicum* B6-PUP-4. Food Technol Biotechnol 2018, 56 (Epub ahead of print).
[37] Sooch BS, Kauldhar BS. Influence of multiple bioprocess parameters on production of lipase from *Pseudomonas sp*. BWS-5. Braz Arch Biol Technol 2013, 56(5), 711–721.
[38] Sooch BS, Kauldhar BS, Puri M. Recent insights into microbial catalases: Isolation, production and purification. Biotechnol Adv 2014, 32(8), 1429–1447.
[39] Kauldhar BS, Sooch BS. Tailoring nutritional and process variables for hyperproduction of catalase from a novel isolated bacterium *Geobacillus* sp. BSS-7. Microb Cell Fact 2016, 17(7), 1–16.
[40] Sooch BS, Lugani, Y. Microbial biodiversity: Types, utility and conservation, In: Chauhan A, Bharti PK, eds. Forest and Biodiversity Conservation. Discovery Publishing House Pvt. Ltd., New Delhi, India, 2017, 131–166.
[41] Kauldhar BS, Dhau JS, Sooch BS. Covalent linkage of alkalothermophilc catalase onto functionalized cellulose. RCS Adv 2016, 6(45), 39364–39375.
[42] Prakash D, Nawani N, Prakash M, Bodas M, Mandal A, Khetmalas M, Kapadnis B. Actinomycetes: A repertory of green catalysts with a potential revenue resource. BioMed Res Int 2013, 2013, 1–8.
[43] Dixit S, Nigam VK. Microbial production of alkaline protease using agricultural by products. Int J Adv Res 2014, 2(6), 407–412.
[44] Sukumaran R, Singhania RR, Pandey A. Microbial cellulases- Production, applications and challenge. J Sci Indus Res 2005, 64, 832–844.
[45] Gurung N, Ray S, Bose S, Rai V. A broader view: Microbial enzymes and their relevance in industries, medicine and beyond. BioMed Res Int 2013, 2013, 1–18.
[46] Maki M, Leung KT, Qin W. The prospects of cellulase producing bacteria for the bioconversion of lignocellulosic biomass. Int J Biol Sci 2009, 5(5), 500–516.
[47] Sooch BS, Kauldhar BS, Puri M. Isolation and polyphasic characterization of a novel hyper catalase producing thermophilic bacterium for the degradation of hydrogen peroxide. Bioprocess Biosyst Eng 2016, 39(11), 1759–73.
[48] Lugani Y, Singla R, Sooch BS. Optimization of cellulase production from newly isolated *Bacillus sp*. Y3. J Bioprocess Biotech 2015, 5(11), 1–6.
[49] Pothiraj C, Balaji P, Eyini M. Enhanced production of cellulases by various fungal cultures in solid state fermentation of cassava waste. Afr J Biotechnol 2006, 5(20), 1882–1885.
[50] Kumar A, Gaind S, Nain L. Evaluation of thermophilic fungal consortium for paddy straw composting. Biodegradation 2008, 19(3), 395–402.
[51] Fadzilah K, Mashitah MD. Cellulases production in palm oil mill effluent: Effect of aeration and agitation. J Appl Sci 2010, 10(24), 3307–3312.
[52] Razak MNA, Ibrahim MF, Yee PL, Hassan MA, Aziz SA. Utilization of oil palm decanter cake for cellulose and polyoses production. Biotechnol Bioproc Eng 2012, 17, 547–555.
[53] Jadhav AR, Chitanand MP, Shete HG. Study on the use of agricultural wastes for cellulose production by using *Aspergillus niger*. IOSR J Pharm Biol Sci 2013, 6(4), 17–20.

[54] Bhoosreddy GL. Comparative study of cellulose production by *Aspergillus niger* and *Trichoderma viride* using solid state fermentation on cellulosic substrates corncob, cane baggase and saw dust. Int J Sci Res 2014, 3(5), 324–326.
[55] Gaur R, Tiwari S. Isolation, production, purification and characterization of an organic solvent thermostable alkalophilic cellulose from *Bacillus vallismortis* RG-07. BMC Biotechnol 2015, 15(19), 1–12.
[56] Siddiqui M, Siddiqui N, Varma R. Production of carboxy methyl cellulose by *Aspergillus niger* using plant wastes as substrate. Int J Curr Microbiol Appl Sci 2015, 4(5), 742–747.
[57] Rahnama N, Shah UKM, Foo HL, Rahman NAA, Ariff AB. Production and characterization of cellulose from solid state fermentation of rice straw by *Trichoderma harzianum* SNRS3. Pertanika J Trop Agric Sci 2016, 39(4), 507–531.
[58] Julia BM, Belen AM, Georgina B, Beatriz F. Potential use of soybean hulls and waste paper as supports in SSF for cellulose production by *Aspergillus niger*. Biocatal Agric Biotechnol 2016, 6, 1–8.
[59] Saheed OK, Jamal P, Karim MIA, Alam MZ, Muyibi SA. Utilization of fruit peels as carbon source for white rot fungi biomass production under submerged state bioconversion. J King Saud Uni Sci 2016, 28(2), 143–151.
[60] Srivastava N, Srivastava M, Mishra PK, Singh P, Ramteke PW. Application of cellulases in biofuels industries: An overview. J Biofuels Bioenergy 2015, 1(1), 55–63.
[61] Singh R, Kumar M, Mittal A, Mehta PK. Microbial cellulases in industrial applications. Annals Appl Bio-Sci 2016, 3(4), 23–29.
[62] Singh RS, Singh RP, Kennedy, JF. Recent insights in enzymatic synthesis of fructooligosaccharides from inulin. Int J Biol Mol 2016, 85, 565–572.
[63] Collins T, Gerday C, Feller G. Xylanases, xylanase families and extremophilic xylanases. FEMS Microbiol Rev 2005, 29(1), 3–23.
[64] Okafor UA, Okochi VI, Okerenta BMO, Chinedu SN. Xylanase production by *Aspergillus niger* ANL 301 using agro wastes. Afr J Biotechnol 2007, 6(14), 1710–1714.
[65] Irfan M, Nadeen M, Syed Q, Baig S. Effect of medium composition on xylanase production by *Bacillus subtilis* using various agricultural wastes. American-Eurasian J Agric Environ Sci 2012, 12(5), 561–565.
[66] Mostafa FA, Aty AAA, Wehaidy HR. Improved xylanase production by mixing low cost wastes and novel co-culture of three marine derived fungi in soil state fermentation. Int J Curr Microbiol Appl Sci 2014, 3(7), 336–349.
[67] Ezejiofor TIN, Enebaku UE, Ogueke C. Waste to wealth- Value recovery from agrofood processing wastes using biotechnology: A Review. Brit Biotechnol J 2014, 4(4), 418–481.
[68] Inacio FD, Ferreira RO, Araujo CA, Peralta RM, Souza CGM. Production of enzymes and biotransformation of orange waste by oyster mushroom, *Pleurotus pulmonarius* (Fr.) Quel Adv Microbiol 2015, 5(1), 1–8.
[69] Olanbiwoninu AA, Odunfa SA. Production of cellulose and xylanase by *Aspergillus terreus* KJ829487 using cassava peels as substrate. Adv Microbiol 2016, 6, 502–511.
[70] Desai DI, Iyer BD. Utilization of corn cob waste for cellulose free xylanase production by *Aspergillus niger* DX-23: Medium optimization and strain improvement. Waste Biomass Volariz 2017, 8(1), 103–113.
[71] Ma Q, Yang R. Alkaline xylanase produced by *Trichoderma reesei*: Application in waste paper pulp bleaching 2015, BioReseource, 10(4), 8048–8057.
[72] Chakda H, Kumar M, Pandiyan K, Singh A, Nanjappan K, Kashyap PL, Srivastava AK. Bacterial xylanases: Biology to biotechnology. 3 Biotech 2016, 6(150), 1–15.
[73] Singh RS, Singh RP. Inulinases. In: Pandey A, Negi S, Soccol CR, eds. Current Developments in Biotechnology and Bioengineering Vol. 7, Elsevier Inc, Cambridge, USA, 2017, 423–46.

[74] Mazutti M, Bender JP, Treichel H, Luccio MD. Optimization of Inulinase production by solid state fermentation using sugarcane bagasse as substrate. Enz Microbiol Technol, 2006, 39, 56–59.
[75] Mazutti M, Ceni G, Luccio MD, Treichel H. Production of inulinase by solid-state fermentation: Effect of process parameters on production and preliminary characterization of *enzyme* preparations. Bioprocess Biosyst Eng, 2007, 30, 297–304.
[76] Dilipkumar M, Rajasimman M, Rajamohan N. Enhanced *inulinase* production by *Streptomyces* sp. in solid state fermentation through statistical designs. 3 Biotech *2013*, 3, 509–515.
[77] Dilipkumar M, Rajasimman M, Rajamohan N. Utilization of copra waste for the solid state fermentative production of inulinase in batch and packed bed reactors. Carbohydr Polym 2014, 102, 261–268.
[78] Ayyachamy M, Khelawan K, Pillay D, Permaul K, Singh S. Production of inulinase by *Xanthomonas campestris* pv *phaseoli* using onion (*Allium* cepa) and garlic (*Allium sativum*) peels in solid state cultivation. Lett Appl Microbiol *2007*, 45, 439–444.
[79] Guo N, Gong F, Chi Z, Sheng J, Li J. Enhanced inulinase production in solid state fermentation by a mutant of the marine yeast *Pichia guilliermondii* using surface response methodology and inulin hydrolysis. J Ind Microbiol Biotechnol, *2009*, 36, 499–507.
[80] Sheng J, Chi Z, Yan K, Wang X, Gong F, Li J. Use of response surface methodology for optimizing process parameters for high inulinase production by the marine yeast *Cryptococcus aureus G7a* in solid-state fermentation and hydrolysis of inulin. Bioprocess Biosyst Eng *2009*, 32, 333–339.
[81] Housseiny MM. Production of an *endoinulinase* from *Aspergillus niger* AUMC 9375, by solid state fermentation of agricultural wastes, with purification and characterization of the free and immobilized enzyme. J Microbiol, 2014, 52, 389–98.
[82] Abd El, Aty AA, Wehaidy HR, Mostafa FA. Optimization of *inulinase* production from low cost substrates using Plackett-Burman and Taguchi methods. Carbohydr Polym 2014, 102, 261–268.
[83] Singh, RS. Enzymatic preparation of high fructose syrup from inulin. In: Panesar PS, Sharma HK, Sarkar BC, eds. Bioprocessing of Foods, Asiatech Publishing Inc, New Delhi, India, 2011, 77–98.
[84] Singh RS, Chauhan K. Production, purification, characterization and applications of fungal inulinases. Curr Biotechnol 2016 (Epub ahead of print)
[85] Singh RS, Chauhan K. Singh RP. Enzymatic approaches for the synthesis of high fructose syrup. In: Gahlawat SK, Salar RK, Siwach P, Duhan JS, Kumar S, Kaur P, eds. Recent Advancements and Developments, Springer, Singapore, 2017a, 189–211.
[86] Singh RS, Chauhan K, Kennedy, JF. A panorama of bacterial inulinases: Production, purification, characterization and industrial applications. Int J Biol Mol 2017b, 96, 312–322.
[87] Singh RS, Singh RP. Production of fructooligosaccharides from inulin by endoinulinases and their probiotic potential. Food Technol Biotechnol 2010, 48, 435–450.
[88] Rafiqul ISM, Sakinah AMM. A Perspective: Bioproduction of xylitol by enzyme technology and future prospects. Int Food Res J 2012, 19(2), 405–408.
[89] Zhang B, Zhang J, Wang D, Gao, X, Sun L, Hong J. Data for rapid ethanol production at elevated temperatures by engineered thermotolerant *Kluyveromyces marxianus* via the NADP (H)-preferring xylose reductase-xylitol dehydrogenase pathway. Data in Brief 2015, 5, 179–186.
[90] Petrova P, Ivanova V. Perspectives for the production of bioethanol from lignocellulosic materials. Biotechnol & Biotechnol EQ 2010, 24, 529–546.

[91] Cadete RM, Melo MA, Dussan KJ, Rodrigues RCLB, Silva SS, Zilli JE, Vital MJS, Gomes FCO, Lachance MA, Rosa CA. Diversity and physiological characterization of D-xylose fermenting yeasts isolated from Brazilian Amazonian forest. Plos One 2012, 7(8), 1–11.
[92] Barathikannan K, Agastian P. Xylitol: Production, optimization and industrial application. Int J Curr Microbiol Appl Sci 2016, 5(9), 324–339.
[93] Kogje A, Ghosalkar A. Xylitol production by *Saccharomyces cerevisiae* overexpressing different xylose reductases using non-detoxified hemicellulosic hydrolysate of corncob. 3 Biotech 2016, 6(2), 1–10.
[94] Tamburini E, Costa S, Marchetti MG, Pedrini P. Optimized production of xylitol from xylose using a hyper-acidophilic *Candida tropicalis*. Biomolecules 2015, 5(3), 1979–1989.
[95] Moyses DN, Reis VC, Almeida JR, Moraes LM, Torres FA. Xylose fermentation by *Saccharomyces cerevisiae*: Challenges and prospects. Int J Mol Sci 2016, 17(3), 1–18.
[96] Lugani Y, Oberoi S, Sooch BS. Xylitol: A sugar substitute for patients of diabetes mellitus. World J Pharm Pharm Sci 2017, 6(4), 741–749.
[97] Lugani Y, Sooch BS. Xylitol, An emerging prebiotic: A review. Int J Appl Pharm Biol Res 2017, 2(2), 67–73.
[98] Rao MB, Aparna M, Tanksale M, Ghatge S, Deshpande VV. Molecular and biotechnological aspects of microbial proteases. Microbiol Mol Biol Rev 1998, 62(3), 597–635.
[99] Vijayaraghavan P, Vincent SGP, Arasu MV, Al-Dhabi NA. Bioconversion of agro-industrial wastes for the production of fibrinolytic enzymes from *Bacillus halodurans* IND 18: Purification and biochemical characterization. Electron J. Biotechnol 2016, 20, 1–8.
[100] Legesse DY. Optimization and partial characterization of *Bacillus* protease isolated from soil and agro-industrial wastes. Int J Nutr Food Sci 2017, 6(1), 31–38.
[101] Sawant R, Nagendran S. Protease: An enzyme with multiple industrial applications. World J Pharm Pharm Sci 2014, 3(6),568–579.
[102] El-Hassayeb A, Aziz SMZ. Screening, production and industrial application of protease enzyme from marine bacteria. Int J Curr Microbiol Appl Sci 2016, 5(7), 863–874.
[103] Sharma N, Rathore M, Sharma M. Microbial pectinase: Sources, characterization and applications. Rev Environ Sci Biotechnol 2013, 12(1), 45–60.
[104] Kohli P, Gupta R. Alkaline pectinases: A review. Biocatal Agric Biotechnol 2015, 4(3), 279–285.
[105] Anisha SK, Girish K. Pectinolytic activity of *Rhizopus sp.* and *Trichoderma viride*. Int J Res Pure Appl Microbiol 2014, 4(2), 28–31.
[106] Garg G, Singh A, Kaur A, Singh R, Kaur J, Mahajan R. Microbial pectinases: An ecofriendly tool of nature for industries. 3 Biotech 2016, 6(1), 1–13.
[107] Ahmed A, Mostafe FA. Utilization of orange baggase and molokhia stalk for production of pectinase enzyme. Braz J Chem Eng 2013, 30(3), 449–456.
[108] Tepe O, Dursan AY. Exo-pectinase production by *Bacillus pumilus* using different agricultural wastes and optimizing od medium components using response surface methodology. Environ Sci Pollut Res 2014, 21(16), 9911–9920.
[109] Biz A, Finkler ATJ, Pitol LO, Medina BS, Krieger N, Mitchell DA. Production of pectinases by solid state fermentation of a mixture of citrus waste and sugarcane baggase in a pilot scale packed bed bioreactor. Biochem Eng J 2016, 111, 54–62.
[110] Ortiz GE, Ponce-Mora MC, Noseda DG, Cazabat G, Saravalli C, Lopez MC, Gill GP, Blasco M, Alberto EO. Pectinase production by *Aspergillus giganteus* in solid state fermentation: Optimization, scale-up, biochemical characterization and its applications in olive oil extraction. J Ind Microbiol Biotechnol 2017, 44(2), 197–211.
[111] Tapre AR, Jain RK. Pectinases: Enzymes for fruit processing industry. Int Food Res J 2014, 21(2), 447–453.

[112] Stambuk P, Tomaskovic D, Tomaz I, Maslov L, Stupic D, Kontic JK. Application of pectinases for recovery of grape seeds phenolics. 3 Biotech 2016, 6(224), 1–12.
[113] Gupta R, Gupta N, Rathi P. Bacterial lipases: An overview of production, purification and biochemical properties. Appl Microbiol Biotechnol 2004, 64(6), 763–781.
[114] Amin M, Bhatti, H. Effect of physicochemical parameters on lipase production by *Penicillium fellutanum* using canola seed oil cake as substrate. Int J Agric Biol, 2014, 16(1), 118–124.
[115] Neethu CS, Rahiman M, Rosmine E, Saramma AV, Hatha AAM. Utilization of agro-industrial wastes for the production of lipase from *Stenotrophomonas maltophilia* isolated from Artic and optimization of physical parameters. Biocatal Agric Biotechnol 2015, 4, 703–709.
[116] Almeida AF, Dias KB, Silva ACC, Terrasan CRF, Tornisielo SMT, Carmona EC. Agroindustrial wastes as alternative for lipase production by *Candida viswanathii* under solid state cultivation: Purification, biochemical properties and its potential for poultry fat hydrolysis. Enzyme Res 2016, 1–15.
[117] Hassan F, Shah AA, Hameed A. Industrial applications of microbial lipasess. Enzyme Microb Tech 2006, 39(2), 235–251.
[118] Thakur S. Lipases, its sources, properties and applications: A review. Int J Sci Eng. Res 2012, 3(7):1–29.
[119] Wang W, Xu Y, Qin X, Lan D, Yang B, Wang Y. Immobilization of lipase SMG1 and its application in synthesis of partial glycerides. Europ J Lipid Sci Technol 2014, 116(8), 1063–1069.
[120] Choudhury P, Bhunia B. Industrial applications of lipase: A review. Biopharm J 2015, 1(2), 41–47.
[121] Fotouh DMA, Bayoumi RA, Hassan MA. Production of thermophilic lipase from *Geobacillus thermovorans* DA2 and application in leather industry. Enzyme Res 2016, 2016, 1–9.
[122] http://www.marketsandmarkets.com [Accessed on 10-4-2017].
[123] www.novozymes.com [Accessed on 10-4-2017].
[124] http://www.specialtyenzymes.com/seb-group-usa [Accessed on 10-4-2017].
[125] http://forbesindia.com/article/breakpoint/novozyme-makes-enzymes-for-abetterlifes /32760/1, 2017 [Accessed on 10-4-2017].

Amit Kumar, Diwakar Aggarwal, Mukesh Yadav, Pawan Kumar and Vikas Kumar

# 3 Biotechnological conversion of plant biomass into value-added products

**Abstract:** Presently, petroleum is undoubtedly the primary feedstock in the production of fuels and several other commodity chemicals. This resource is not only finite but also depleting rapidly. Hence, te transformation of plant biomass into high-value-added products is a suitable alternative to petroleum-based fuels and chemicals. Plant biomass mainly consists of three polymers: cellulose, hemicelluloses and lignin. The biotechnological conversion of plant biomass comprises three steps: pretreatment, hydrolysis of the pretreated biomass and fermentation of reducing sugars into fuels and biochemicals. Numerous high-value-added products such as bioethanol, 1-butanol, bio-methane, bio-hydrogen, citric acid, fumaric acid, lactic acid, succinic acid, single cell protein, curdlan, dextran, schizoplyllan, pullulan, xylitol and other fine chemicals can be obtained from plant biomass.

**Keywords:** Plant biomass, value-added products, biofuels, microbial polysaccharide, citric acid, enzymes

## 3.1 Introduction

Presently, petroleum is undoubtedly the primary feedstock in the production of fuels and several other commodity chemicals. This resource is not only finite but also depleting rapidly. The extreme use of fossil fuels has resulted in environmental pollution because of the release of greenhouse gases. With the development in plant-biomass-conversion technologies, plant biomass is now a potential alternative for petroleum refinery [1, 2]. Every year, a vast amount of residues of plant biomass is generated throughout the world, which, if released in the environment, may create several environmental problems [3]. Agricultural residues, industrial wastes and forest plant biomass are feedstock that can be used for sustainable generation of biofuels and biochemical products. Plant biomass, which is a major component of biomass available on earth, is renewable and abundantly available, with a great potential for bioconversion to value-added bio-products. Plant biomass is mainly composed of cellulose, hemicelluloses and lignin biopolymers [4]. Industrial products such as bioethanol, 1-butanol, bio-methane, bio-hydrogen, organic acids, microbial polysaccharides, single-cell protein and xylitol and other fine chemicals can be obtained from lignocellulosic plant biomass (Figure 3.1). The process of bioconversion of lignocellulosic biomass comprises three steps. In the first step, lignocellulosic biomass is pretreated to enhance the accessibility of carbohydrates by the removal of lignin.

https://doi.org/10.1515/9783110563337-003

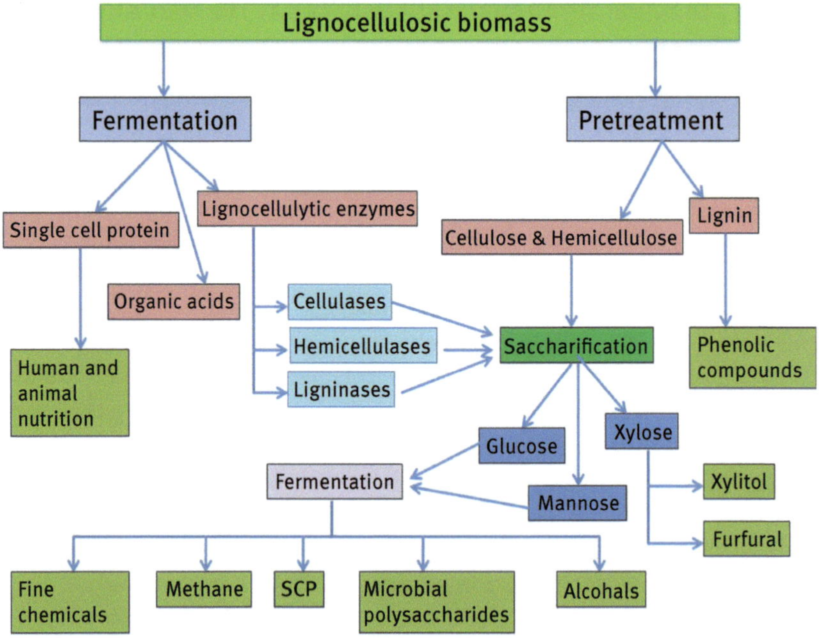


Figure 3.1: Conversion of plant biomass into value-added products (adapted with permission from ref.[6]).

In the second step, lignocellulosic biomass is hydrolyzed into fermentable reducing sugars by means of acid hydrolysis or enzymatic hydrolysis. A variety of microorganisms including bacteria and fungi may produce cellulose- and hemicellulose-degrading enzymes and have the ability to degrade the cellulose and hemicelluloses into monomer sugars [5, 6]. In the third step, fermentable reducing sugars are metabolized by different microorganisms into the desired biochemicals [4].

## 3.2 Lignocellulosic biomass composition

The major components of lignocellulosic biomass are cellulose, hemicelluloses, and lignin. Cellulose and hemicelluloses are carbohydrates made up of different sugars, and lignin is an aromatic polymer of phenyl propane units. Cellulose (30–60%), hemicelluloses (20–40%) and lignin (10–25%) are intermeshed with one another in lignocellulosic material. Cellulose is a linear polymer that is composed of D-glucose units which are linked by β-glycosidic linkage. Hemicelluloses are the polymers of pentose and hexose sugars. D-xylose and L-arabinose are the pentose sugars that are present in hemicelluloses, whereas D-mannose, D-glucose and D-galactose are the hexose sugars that are found in hemicelluloses. The relative contents of cellulose, hemicelluloses and lignin depend on the type of plant biomass and vary in different lignocellulosic materials [6, 7] (Table 3.1).

Table 3.1: Chemical composition of several plant biomasses (adapted with permission from ref. [6]).

| Plant biomass | Lignin (%) | Hemicelluloses (%) | Cellulose (%) | References |
|---|---|---|---|---|
| Bamboo | 21–31 | 15–26 | 26–43 | [7, 8] |
| Banana waste | 14 | 14.8 | 13.2 | [7] |
| Barley straw | 14–15 | 24–29 | 31–34 | [8, 9] |
| Bast fiber jute | 21–26 | 18–21 | 45-–3 | [7, 8] |
| Bast fiber kenaf | 15–19 | 22–23 | 31–39 | [7, 8] |
| Black gram residue | 23.14 | 32.48 | 26.80 | [10] |
| Coastal Bermuda grass | 6.4<br>19.4 | 35.7<br>13.3 | 25<br>47.8 | [11, 12]<br>[13] |
| Coconut coir | 18 | 26 | 48 | [14] |
| Corncob | 15<br>4.5–6.6 | 35<br>38–42 | 45<br>35 | [7, 11, 12]<br>[15] |
| Corn stalks | 17 | 24 | 43 | [14] |
| Corn stover | 18<br>18 | 22<br>20.5 | 40<br>41.7 | [16]<br>[17] |
| Cotton gin waste | – | 16 | 78 | [14] |
| Elephant grass | 24 | 24 | 22 | [7] |
| Esparto grass | 17–19 | 27–32 | 33–38 | [7, 8] |
| Flax straw | 22 | 27 | 29 | [9] |
| Grasses (average) | 10–30 | 25–50 | 25–40 | [18] |
| Hardwood stem | 18–25 | 24–40 | 40–55 | [7, 11, 12, 18] |
| Leaves | 0 | 80–85 | 15–20 | [11, 12] |
| Millet husk | 14 | 27 | 33 | [14] |
| Newspaper | 18–30 | 25–40 | 40–55 | [11, 12] |
| Nut shells | 30–40 | 25–30 | 25–30 | [11, 12] |
| Oat straw | 16–19 | 27–38 | 31–37 | [7, 8] |
| Orchard grass (medium maturity) | 4.7 | 40 | 32 | [7, 11] |
| Pinewood | 20 | 24 | 39 | [19] |

Table 3.1 (continued)

| Plant biomass | Lignin (%) | Hemicelluloses (%) | Cellulose (%) | References |
|---|---|---|---|---|
| Poplar wood | 26 | 17 | 35 | [20] |
| Rice husk | 14 | 24 | 31 | [14] |
| Rice straw | 18 | 24 | 32.1 | [7, 12, 14] |
| Rye straw | 16–19 | 27–30 | 33–35 | [7, 8] |
| Sabai grass | 20.88 | 23.72 | 49.90 | [21] |
| Softwood stem | 25–35 | 25–35 | 45–50 | [7, 11, 18] |
| Sugarcane bagasse | 19–24 | 27–32 | 32–44 | [8, 14] |
| | 18.4 | 26.9 | 38.1 | [22] |
| | 24 | 25 | 43 | [17] |
| Sugarcane tops | 36.1 | 24.2 | 33.3 | [23] |
| Sunhemp residue | 17.4–18.4 | 11.9–13 | 43.4– 48 | [24] |
| Sweet sorghum bagasse | 18 | 25 | 45 | [20] |
| Switchgrass | 12 | 31 | 45 | [7] |
| | 18 | 22 | 31 | [17] |
| Timothy grass | 18 | 30 | 34 | [19] |
| Waste papers from chemical pulps | 5–10 | 10–20 | 60–70 | [11, 12] |
| Wheat bran | 3–10 | 22–25 | 7–11 | [25] |
| Wheat straw | 16–21 | 26–32 | 29–35 | [11] |
| | 7.7 | 30.8 | 41.3 | [26] |
| | 15 | 50 | 30 | [11] |
| | 17 | 23 | 33 | [17] |
| | 14.5 | 24.8 | 36 | [22] |

## 3.3 Pretreatment of lignocellulosic biomass

Lignocellulosic biomass is recalcitrant to microbial attack and enzymatic hydrolysis due to the presence of lignin. Factors such as lignin content, cellulose accessibility to cellulase and cellulose crystallinity decide the overall digestibility of a lignocellulosic material. Therefore, lignocellulosic biomass must be subjected to pretreatment to remove the lignin and enhance the digestibility of cellulose and hemicelluloses, making it suitable for enzymatic hydrolysis. Several key features are involved in the effective pretreatment of lignocellulosic biomass. The capital and operational cost should be low for a successful pre-treatment. The pretreatment should be applied on a wide range of lignocellulosic biomass and should be able to recover most of the lignocellulosic components in a useable form. Lignin degradation products that inhibit the growth of fermentative microorganisms or the action of hydrolytic enzymes should not be generated [27, 28].Various pretreatment methods have been tested and are still in development. The different pretreatment methods can be classified into physical, chemical, physicochemical, biological and multiple or combinatorial pretreatments. Evaluation and comparison of different pretreatment processes are not easy because they involve upstream and downstream processing costs, capital investment, chemical recycling and waste treatment systems [27, 29]. Physical pretreatment aims at reducing particle size and crystallinity using milling or grinding. The final particle size and reduction in crystallinity of the lignocellulosic biomass determine the requirements of energy and cost for physical pretreatment. Several types of physical pretreatments, such as ball milling, two-roll milling, hammer milling, colloid milling and vibro-energy milling, have been utilized for pretreatment of lignocellulosic biomass. Different types of irradiations such as by gamma rays, electron beam and microwaves can also be used to enhance the enzymatic saccharification of plant biomass. The combination of both physical and chemical processes is referred to as the physicochemical process [30]. Various physicochemical processes such as steam explosion, ammonia fiber explosion, $CO_2$ explosion, liquid hot water pretreatment and microwave-chemical pretreatment, have been used successfully. Some chemicals such as acids, alkali, organic solvents and ionic liquids have been reported to have significant effects on the native structure of lignocellulosic biomass. Biological pretreatments utilize the action of fungi enzymes that can degrade lignin, hemicelluloses and polyphenols. White-rot fungi and soft-rot fungi are mainly used for the biological pretreatment of biomass [27].

## 3.3.1 Hydrolysis of pretreated biomass

Hydrolysis of cellulose and hemicelluloses is one of the key steps in the generation of value-added products. Carbohydrates available in the plant biomass can be hydrolyzed into fermentable sugars by either acid hydrolysis or enzymatic action. Acid hydrolysis is a relatively simple process that utilizes dilute or concentrated acid. Several acids such as sulfuric, sulfurous, hydrochloric, hydrofluoric, phosphoric, nitric and formic acids have been utilized for lignocellulosic biomass hydrolysis. Concentrated acid hydrolysis is performed in the 10–30% range at low temperatures, which produces a higher yield of fermentable reducing sugars. The main disadvantages of concentrated acid hydrolysis include the strong equipment corrosion and inadequate acid recovery. This process requires large amounts of acids. Although dilute acid hydrolysis requires a low amount of acid (2–5%), it has to be performed at high temperature and pressure to achieve an acceptable conversion of carbohydrates. The high temperature required during hydrolysis results in the decomposition of hemicellulose sugars that generates toxic compounds, including furfural and 5-hydroxymethyl-furfural. These compounds act as inhibitors during fermentation and lower the ethanol production by yeasts. Therefore, acid hydrolysis methods are unsuitable at the industrial scale due to high equipment cost and environmental issues. Enzymatic hydrolysis is a relatively complex but better alternative for the hydrolysis of lignocellulosic biomasses. The complex structures of lignin, hemicelluloses and cellulose prevent the effective hydrolysis of carbohydrates; thus, some pretreatments are required before enzymatic hydrolysis. Enzymatic hydrolysis has been proved to have several advantages such as high yield of fermentable reducing sugars, mild reaction conditions and negligible by-product formation. The high cost of enzymes and the requirement of an extra step of pretreatment limit their use at the industrial scale [31–33]. Simple reducing sugars can be liberated effectively from celluloses and hemicelluloses using appropriate enzymes. Hydrolysis of cellulose is performed with three classes of enzymes, namely endoglucanase, exoglucanase and β-glucosidase, collectively called cellulases. Xylan is the main component of hemicelluloses, and it is hydrolyzed by endoxylanases and β-xylosidases. Enzymatic hydrolysis is influenced by several enzyme-related and substrate-related factors. Substrate-related physical and chemical factors include lignin content, the extent of lignin and xylan removal during pretreatment, lignin structure, particle size of substrate, porosity and accessibility of substrate, and crystallinity of cellulose [34, 35]. Enzyme-related factors include loading of enzymes, synergy of enzymes, generation of inhibitors during pretreatment, addition of surfactant, and enzyme inhibition by monosaccharides, disaccharides and oligosaccharides. Hydrolysis reaction conditions such as temperature, pH and agitation are the factors that influence the enzymatic hydrolysis of lignocellulosic biomass. These reaction conditions must be optimized to attain maximum hydrolysis of lignocellulosic biomass [36].

### 3.3.2 Fermentation of reducing sugars

After hydrolysis, reducing sugars such as glucose and xylose are fermented by microbes into various value-added products such as ethanol, butanol, methane, hydrogen, citric acid, succinic acid and lactic acid, microbial polysaccharides and xylitol. Bioenergy products are produced using various strategies including separate hydrolysis and fermentation, simultaneous saccharification and fermentation, simultaneous saccharification and co-fermentation, and consolidated biomass processing (CBP). The integration of steps reduces the capital cost of the process [37].

## 3.4 Plant biomass conversion to bioenergy products

Globally, energy demand is increasing rapidly, and energy consumption in the world may reach up to 240 quadrillion in 2040 according to statistics of US Energy Information Administration. Fossil fuels are depleting continuously, and hence they cannot fulfill the increasing energy demands. Therefore, efficient, sustainable and renewable potential alternatives are required to supply the fuels in a sustainable and eco-friendly way. Plant biomass, which is abundantly available in nature, is a potential resource for biotechnological conversion into different bioenergy products such as bioethanol, biobutanol and biohydrogen. Monomeric sugars can be obtained from pretreated plant biomass by enzymatic hydrolysis efficiently. Monomeric sugars can be fermented into various bioenergy products using different microorganisms [6, 38].

### 3.4.1 Bioethanol production

Commercially, ethanol is produced from sugars and starch found in food crops such as sugarcane, maize or potato. These are called first-generation biofuels, which compete with human food supply. Lignocellulosic materials, on the other hand, are abundant and a renewable resource for ethanol production, and they do not compete with food crops [39]. Bioethanol production from lignocellulosic materials could be a sustainable alternative to reduce the need of fossil fuels. The increasing fuel demand has made efficient ethanol production from lignocellulosic biomass an issue of intensive research [40]. Cellulosic bioethanol is a potential near-term technological and sustainable alternative that can produce a larger proportion of the current supply of ethanol to reduce greenhouse gas emissions in the transportation sector [39]. Several plant biomasses like Bermuda grass, corncob, empty fruit bunches of oil palm, rapeseed stover, reed residue, rice husk, rice straw, sugarcane bagasse, sunflower hulls, sunflower stalks, sweet sorghum residue, water hyacinth residue and wheat straw have been tested for ethanol production [6]. An ideal organism for ethanol production is one that can produce high ethanol yield, should be tolerant to high ethanol concentration to minimize cost of the distillation process and should be able to withstand higher concentrations of inhibitors as well as higher temperatures [2, 41]. *Saccharomyces cerevisiae* is the most suitable microorganism for the production of ethanol due to several reasons. It has a higher yield of ethanol production, along with higher ethanol and substrate tolerance [2]. *S. cerevisiae* can metabolize glucose, fructose, galactose, maltose and sucrose into ethanol, but it cannot metabolize pentoses, lactose and cellobiose. Plant biomasses are rich hemicelluloses, and their saccharification produces a large amount of xylose, a pentose sugar that cannot be utilized by *S. cerevisiae* for ethanol production. Pentoses such as xylose can be easily converted into ethanol by some other yeasts including *Pichia stipitis, P. segobiensis, Candida shehatae, Pachysolen tannophilus* and *Hansenula polymorpha*. *Pichia stipitis* can ferment xylose very efficiently. It can also ferment other reducing sugars such as glucose, galactose and cellobiose under anaerobic conditions. Furthermore, the ethanol resistance of *Pichia stipitis* is lower compared to that of *S. cerevisiae*. *Zymomonas mobilis* is a Gram-negative, facultative anaerobic bacterium that can ferment glucose, fructose and sucrose but cannot ferment pentose sugars like xylose and arabinose, which are produced in large quantities during the saccharification of lignocellulosic materials [42].

### 3.4.2 Biobutanol production

Biobutanol is produced through acetone–butanol–ethanol (ABE) fermentation by several species of *Clostridium*. Butanol is an emerging biofuel and biochemical that has the potential to replace gasoline. It is an alcohol with interesting properties such as engine compatibility, good safety, and energy content and octane number close to gasoline. Butanol is also used as a raw material for manufacturing butyl phthalate, butyl acrylate and glycerol ethers in the chemical industry [6, 43]. Traditionally, butanol is manufactured from crotonaldehyde dehydrogenation and hydroformilation of olefins, which are derived from crude oil. Biotechnologically, it is produced through ABE fermentation by different species of *Clostridium* [43]. Annually, approximately US$15–17 B of butanol is generated globally [44]. Recent studies have focused on developments in the industry to enhance the efficiency of ABE fermentation to make it cost effective. The future of biobutanol production depends on the utilization of inexpensive lignocellulosic substrates and hyper-butanol producing bacterial strains. Substrate selection and conditioning are important for the production of biobutanol because they account for approximately 60% of the total manufacturing cost. Feedstock is selected for ABE fermentation based on its abundance, annual supply and processing cost [45]. Several lignocellulosic substrates have been studied for biobutanol production (Table 3.2).

### 3.4.3 Biomethane production

Biomethane has the potential to yield more energy than any other current biofuels like bioethanol and biodiesel. Biogas is used for cooking purposes, electricity generation and as vehicle fuel where biogas is cleaned and upgraded to biomethane, which is combusted in an internal combustion engine. Anaerobic digestion (AD) technology, which is an efficient process for the conversion of organic waste into energy, is a microbiological process that converts organic materials into biogas in the absence of oxygen. Biogas chiefly consists of methane and carbon dioxide [6]. AD is a three-step

Table 3.2: Bioenergy products produced from plant biomass by microbial conversion.

| Plant biomass | Pretreatment | Source for hydrolysis | Fermentation microorganism | Ref. |
|---|---|---|---|---|
| **Bioethanol production** | | | | |
| Sugarcane straw | $H_2SO_4$ treatment at high temperature | Cellic®CTec2 VCN10005 (Novozymes) | *Saccharomyces cerevisiae* no. 1701 | [50] |
| *Paulownia tomentosa biomass* | Non-isothermal autohydrolysis | Celluclast 1.5 L cellulases, Novozyme 188-β-glucosidase (Novozymes) | *Saccharomyces cerevisiae* CECT-1170 | [51] |
| Olive tree biomass | Sequential acid/ alkaline oxidative | Cellic Ctec 3-cellulase & Novozyme 50010-β-glucosidase (Novozymes) | *Escherichia coli* MM160 | [52] |
| Sugarcane bagasse | Ammonia & dilute acid pretreatment | Xylanase & β-glucosidase by *Aspergillus niger* ADH-11, commercial cellulase (MAPs 450) | *Saccharomyces cerevisiae* NCIM 3521 &*Pichia stipitis* NCIM 3506 | [40] |
| Wheat straw | Steam explosion | NS50013 (Celluclast 1.5 L) & NS50010 (Novozyme 188), Novozymes | *Kluyveromyces marxinus* CECT 10875 | [53] |
| *Saccharum spontaneum* biomass | Dilute sulfuric acid, sodium hydroxide and aqueous ammonia treatment | *Aspergillus oryzae* MTCC1846 | *Pichia stipitis* NCIM3498 | [54] |
| Rice straw | Acidic, alkaline and hydrogen peroxide pretreatment | *Trichoderma reesei* SAP3 MTCC4876 | *Saccharomyces cerevisiae* MTCC 173 & *Zymomonas mobilis* MTCC 2428 | [55] |
| Bermudagrass, reed & rapeseed stover | Phosphoric acid-acetone pretreatment | Celluclast 1.5 L (Novozymes) | *Saccharomyces cerevisiae* ATCC #24858 | [13] |
| **Acetone–Butanol–Ethanol (ABE) Production** | | | | |
| Wheat straw | Acid treatment | – | *Clostridium beijerinckii* | [56] |

Table 3.2 (continued)

| Plant biomass | Pretreatment | Source for hydrolysis | Fermentation microorganism | Ref. |
|---|---|---|---|---|
| Corn stover/ Switchgrass | Dilute acid with high temperature | Celluclast 1.5 L & Novozyme 188 (Sigma Chemicals) & Viscostar 150L (Dyadic Corporation, Jupiter) | *Clostridium beijerinckii* P260 | [57] |
| Corncob | NaOH pretreatment | Cellic®CTec2 (Novozymes) | *Clostridium saccharobutylicum* DSM13864 | [58] |
| King grass | Alkaline delignification | Acellerase 1500 (Genencor) | *Clostridium acetobutylicum* ATCC 824 | [43] |
| Rice straw | | Celluclast 1.5 L &Novozyme 188 (Novoenzymes) | *Clostridium acetobutylicum* NRRL B-591 | [59] |
| **Methane production** | | | | |
| Rice straw | Extrusion with alkaline pretreatment | – | Anaerobic digestion | [60] |
| Maize straw, grass, sprout stem | Extrusion with $(CaOH)_2$ treatment | – | Anaerobic digestion | [61] |
| *Agave tequilana* bagasse | Mechanical size reduction | Dilute acid hydrolysis | Anaerobic digestion | [62] |
| *Pennisetum hybrid* | Alkaline pretreatment | – | Anaerobic digestion | [63] |
| Rapeseed leaves & stem | Rumen fluid treatment | – | Anaerobic digestion | [64] |
| Switch grass | Alkaline pretreatment | Celluclast 1.5 L & Novozyme 188 (Novoenzymes) | Anaerobic digestion | [65] |
| **Hydrogen** | | | | |
| Sweet sorghum bagasse | Alkaline pretreatment | GC 220 Genencor International | *Caldicellulosiruotor saccharolyticus* | [66] |
| Pine tree wood | Dilute acid pretreatment | – | Anaerobic digestion | [49] |

Table 3.2 (continued)

| Plant biomass | Pretreatment | Source for hydrolysis | Fermentation microorganism | Ref. |
|---|---|---|---|---|
| Rice straw | Ethanol organosolv pretreatment | Cellic®CTec2 (Novozymes) | *Enterobacter aerogenes* | [47] |
| Wheat straw | Dilute acid treatment | Dilute acid hydrolysis | *Escherichia coli* WDHL | [67] |
| Corn stalk | Fungal (*Trichoderma reesei* Rut C-30) pretreatment | – | Thermophilic anaerobic digestion | [68] |
| Wheat bran | Acidic & microwave-assisted acidic pretreatment | – | Anaerobic digestion | [69] |
| Sugarcane bagasse | Size reduction | Acid hydrolysis | *Thermoanaerobacterium aotearoense* SCUT27/Δ/dh | [70] |
| Sunflower stalk | Thermochemical pretreatments | | Anaerobic digestion | [71] |

process consisting of hydrolysis, acidogenesis and methanogenesis. Biomethane production from lignocellulosic biomass is a better option than bioethanol production due to the maximum conversion of the biomass content (carbohydrates, fats and proteins) during the AD process. These biological macromolecules are first converted into simple derivatives and finally into methane and carbon dioxide by several kinds of anaerobic and methanogenic bacteria [46]. Different lignocellulosic biomasses have shown potential for conversion into methane (Table 3.2).

### 3.4.4 Biohydrogen production

Hydrogen is the cleanest alternative fuel with the highest energy content. Traditionally, hydrogen is generated from natural gas. Biohydrogen production from plant biomasses is a renewable approach to meet the global hydrogen demand for utilization as fuel or nitrogen fertilizer manufacturing [47]. Zero greenhouse gas emission is achieved using hydrogen as a fuel. Biohydrogen is produced via a dark fermentation or light-independent process that is performed by *Clostridium*, *Enterobacter* or mixed cultures. *Clostridium* species have been proved the most dominant producer of hydrogen under mesophilic dark fermentation at pH 5.5.

The species of genus *Clostridium* metabolize the fermentable sugars available in the dilute hydrolyzate of lignocellulosic biomass [48, 49]. Different plant residues such as barley straw, rice straw, corn stalk, corncob, wheat bran, wheat straw, sugarcane bagasse, sweet sorghum bagasse, hemp and newspaper have been utilized for the production of hydrogen [6].

## 3.5 Production of industrial enzymes using plant residues as the substrate

Enzymes are biocatalysts that transform macromolecules to energy and new materials, which besides other purposes may be needed for the growth, repair and maintenance of cells [72]. Enzymes have shown great potential for several industrial processes. A major issue with the industrial application of enzymes is the cost of enzymes. Large-scale production in the industrial scale is an expensive process, and use of enzymes during the manufacturing process determines the cost of the final products. Plant equipment, installation and raw materials all contribute to the cost of enzyme production. Lignocellulosic biomasses are cheap and abundantly available raw materials that can act as substrates for the production of enzymes such as xylanases, cellulases, lignases, pectinases, proteases and amylases [73]. Black gram residue, corn bran, corncob, corn stover, *Eichhornia crassipes* waste, pearl millet stover, rice straw, rice husk, soybean bran, sweet sorghum, sugarcane bagasse, Switch grass, *Parthenium hysterophorous* biomass, wheat bran, wheat straw and other agro-industrial residues have been used for the production of enzymes (Table 3.3). Enzymes are produced from microorganisms like bacteria, fungi and actinomycetes using solid-state fermentation or submerged fermentation using agro-industrial residues cost effectively [6].

## 3.6 Organic acid production

Organic acids are the building-block and platform chemicals produced by microorganisms. These are natural products of microorganisms or intermediates in important metabolic pathways. Because of their functional group, organic acids are very useful starting materials for the synthesis of other chemicals [93]. Currently, the chemical industry is focusing on potential high-value organic acids with broad industrial applications. Several organic acids such as succinic, fumaric, propionic itaconic and lactobionic acids have proved potential platform chemicals for the synthesis of several fine chemicals including pharmaceuticals and food additives [94]. Natural production of organic acids by fungi, which have been proved as efficient producers of organic acids, plays key roles in nature depending on the species

Table 3.3: Plant biomass utilization for conversion into industrial enzymes by microbes.

| Plant biomass | Microorganism | Enzyme | Reference |
|---|---|---|---|
| Wheat bran | *Rhizopus microsporus var. oligosporus* | Amylase | [74] |
| Corn bran | *Bacillius amyloliquefaciens* NRRL B-645 | Amylase | [75] |
| Wheat bran | *Rhizopus oligosporus* | Amylase | [76] |
| Wheat bran | *Aspergillus fumigatus* NTCC1222 | Amylase | [77] |
| Soybean bran | *Aspergillus oryzae, Penicillium roquefortii* and *Aspergillus flavipes* | Protease | [78] |
| Black gram residue | *Aspergillus nidulans* AKB-25 | Cellulase, xylanase, β-glucosidase | [79] |
| Pearl millet stover | *Aspergillus flavus* ARC-12 | Xylanase | [80] |
| Wheat straw | *Aspergillus tubingensis* JP-1 | Xylanase | [81] |
| Rice straw | *Schizophyllum commune* ARC-11 | Xylanase | [82] |
| Wheat bran | *Trichoderma harzianum* PPDDN10 NFCCI-2925 | Cellulase, xylanase | [83] |
| Wheat bran | *Aspergillus niger* NFCCI 4113 | Cellulase, xylanase, β-glucosidase | [84] |
| Sugarcane bagasse/ wheat bran | *Myceliophthora heterothallica* | Endoglucanase, β-glucosidase | [85] |
| *Parthenium hysterophorous* biomass | *Talaromyces stipitatus* MTCC 12687 | Cellulase, xylanase, β-glucosidase | [86] |
| Wheat bran | *Penicilliun*sp. AKB-24 | Cellulase, xylanase, β-glucosidase | [87] |
| Switch grass | *Trichoderma reesei* QM6a, *Trichoderma asperellum* T-1 | Cellulases, xylanase, β-glucosidase | [88] |
| Sweet sorghum | *Trichoderma harzianum* HZN-11 | Cellulases, xylanase, β-glucosidase | [89] |
| Rice straw & husk, sunflower seed hulls | *Ganoderma lucidum* | Laccase | [90] |

Table 3.3 (continued)

| Plant biomass | Microorganism | Enzyme | Reference |
|---|---|---|---|
| Oil palm, banana and sugarcane residues | *Pleurotus sajar-caju* | Laccase, manganese peroxidase | [91] |
| *Eichhornia crassipes* waste | *Pycnoporus sanguineus SYBC-L1* | Laccase | [92] |

of fungi producing them [95]. Fermentative production of organic acids has gained a key position in the chemical industry due to its cost effectiveness. Microbial production of metabolite via fermentation uses costly media that accounts for 60% of the production cost. This led to the use of inexpensive substrates from nonfood renewable resource to make the process profitable and sustainable [94].

Citric acid –one of the most common transitional metabolic products –is widely used in the food, beverage and pharmaceutical industries. It is an environmentally safe compound, and it is used for chelation, buffering, pH adjustment and derivatization. It is also used in laundry detergents, shampoos and cosmetics and for enhanced oil recovery as well as chemical cleaning [96]. Globally, citric acid is the second largest fermentation product after ethanol, with 1.7 million tons of production annually. Citric acid is dominantly produced under submersed and solid-state fermentation by several species of genera *Aspergillus* and *Penicillium*. Some bacteria are also found to produce citric acid, in addition to several plant- and fruit-based wastes such as apple pomace, cassava bagasse, cotton waste, corncobs, corn husk, oil palm empty fruit bunches, pineapple waste and sugarcane bagasse [6].

Fumaric acid, which is white in color, has a low aqueous solubility. Previously, fumaric acid was produced from petroleum by a chemical process. Microbial fermentation produces fumaric acid using biomass as the substrate. It is a four-carbon dicarboxylic acid that is 1.5 times more acidic than citric acid. Therefore, it is used as acidulant in the food industry and as a beverage ingredient. It also acts an antibacterial agent for the feed industry. Fumaric acid also has the ability to transform into synthetic resins and biodegradable polymers. The zygomycetes fungus *Rhizopus oryzae* is a well-proven and efficient producer of fumaric acid. The fermentation cost for fumaric production can be minimized using renewable raw materials such as molasses and starch. Plant biomass is a potential alternative of starch for fumaric production [95, 97]. Woiciechowski *et al.* studied the production of fumaric acid by *Rhizopus* species using steam-exploded wood chip hydrolysate as a substrate [98].

Succinic acid is a four-carbon dicarboxylic acid that is listed in the US Department of Energy's top 12 bio-based molecules. It not only has wide applications in the food, pharmaceutical and agriculture industries, but also is utilized for biodegradable plastics, detergents, green solvents and surfactants. Traditionally, succinic acid is produced from petroleum-derived precursors. Microbial production of succinic acid is an alternative pathway that involves anaerobic bacterial fermentation of pure sugars. Plants biomass can be used as a substrate for succinic acid production but requires pretreatment and hydrolysis to liberate reducing sugars that can be metabolized by succinic acid-producing bacteria [6, 99]. Currently, some of the renewable plant residues used for succinic acid production are *Cannabis sativa* L, corn stover, corn stalk hydrolysate, pinewood, sugarcane bagasse hydrolysate and wood hydrolysate [6].

Lactic acid is a versatile organic acid that has been used as an emulsifying, acidulant, buffering and preservative agent in the food industry. It has also been applied in textile, detergent and paper industries due to its antibacterial activity and ink-erasing ability. Succinic acid also finds important applications in the pharmaceutical, cosmetic and chemical industries. The global lactic acid market was 714 kt in 2013, and it might reach up to 1960 kt by 2020. Lactic acid finds new application in the plastic industry, where polylactic acid is used as biodegradable plastic to decrease the net emission of $CO_2$ and demand for petroleum [6, 100, 101]. Lactic acid can be produced either by chemical synthesis or by microbial fermentation. During chemical synthesis, lactic acid is derived from petroleum, which has a negative impact on the environment. Moreover, racemic mixture is produced via chemical synthesis. Contrarily, optically pure lactic acid is produced by microbial fermentation. Lactic acid is dominantly produced by different species of the genus Lactobacillus. Globally, most of the lactic acid is produced via fermentation using starch as the substrate. However, utilization of starch competes with food supply. Therefore, alternative nonfood raw materials are required as a substitute for starch-based lactic acid production. The use of fermentable sugars from plant biomass is a suitable and sustainable alternative [100]. Plant residues including corn stover, corncob, office waste paper, paper sludge, rice bran, sugarcane bagasse, and wheat bran have been tested for the production of lactic acid [6].

## 3.7 Conversion of plant biomass into other value-added products

Several high-value-added products, such as microbial polysaccharide, single-cell proteins and xylitol, can be generated from plant residues. Polysaccharides are used as thickeners, stabilizers and gelling agents in the food industry. They are also useful as depollution compounds, antitumor agents, antioxidants and prebiotic agents. Different microbial bioactive polysaccharides such as schizophyllan, pullulan, xanthan, dextran and curdlan are produced by microorganisms using fermentable sugars as the substrate. Hydrolysates from plant residues are the potential sources of reducing sugars that are fermented into polysaccharides by microorganisms [6, 102]. Schizophyllan is a homoglucan that is composed of a β-1,3-linked backbone and single β-1,6-linked glucose side chains at every third residue of the backbone. Timothy *et al.*(2016)[103] reported the production of schizophyllan by white-rot fungus *Schizophyllum commune* using corn fiber as the substrate. Pullulan is an extracellular polysaccharide produced by *Aureobasidium* spp., which is a linear macromolecule linked by α-1,6-glycosidic linkage of maltotrise repeating units. It is highly water soluble and biodegradable due to its unique structure. Pullulan is widely applicable in food processing and preservation. It is also used as adhesive and in the cosmetics industry [104, 105].

Xylitol ($C_5H_{12}O_5$) is a sugar alcohol that is used as a natural food sweetener with low caloric value. It has similar sweetening power as sucrose and is very useful for diabetic persons. The use of xylitol is also beneficial for oral health, because it can reduce tooth decay. Traditionally, xylitol is produced by the chemical hydrogenation of xylose in the presence of a metal catalyst under high temperature and pressure. It is an energy-intensive and costly process. Therefore, microbial fermentation of xylitol production using xylose available in hemicellulose hydrolysate is an economic and attractive alternative [105, 107]. Although several bacteria, fungi and yeasts have the capability to metabolize xylose for the production of xylitol, yeasts are the most efficient producers of xylitol. Xylose is the sugar that is released from the hydrolysis of hemicelluloses. The use of agro-industrial wastes has been widely investigated for their potential use in xylitol production [6, 108].

## 3.8 Conclusions

Currently, future trends have been directed toward lignocellulose biotechnology for eco-friendly and sustainable processes. Plant biomass has shown potential for the replacement of petroleum for the generation of fuels and biochemicals. Energy and environmental crises have led to the need to develop efficient technologies for the conversion of plant residues into valuable products.

## References

[1] Gupta A, Verma JP. Sustainable bio-ethanol production from agro-residues: A review. Renew Sustain Energy Rev 2015, 41, 550–67.

[2] Adsul MG, Singhvi MS, Gaikaiwari SA, Gokhale D V. Development of biocatalysts for production of commodity chemicals from lignocellulosic biomass. Bioresour Technol 2011, 102, 4304–12.

[3] Sadh PK, Duhan S, Duhan JS. Agro-industrial wastes and their utilization using solid state fermentation: a review. Bioresour Bioprocess 2018, 5, 1–15.

[4] Hasunuma T, Okazaki F, Okai N, Hara KY, Ishii J, Kondo A. A review of enzymes and microbes for lignocellulosic biorefinery and the possibility of their application to consolidated bioprocessing technology. Bioresour Technol 2013, 135, 513–22.
[5] Kumar R, Singh S, Singh OV. Bioconversion of lignocellulosic biomass: biochemical and molecular perspectives. J Ind Microbiol Biotechnolgy 2008, 35, 377–91.
[6] Kumar A, Gautam A, Dutt D. Biotechnological transformation of lignocellulosic biomass in to industrial products: An overview. Adv Biosci Biotechnol,2016, 7, 149–68.
[7] Sanchez C. Lignocellulosic residues: Biodegradation and bioconversion by fungi. Biotechnol Adv 2009, 27, 185–94.
[8] Rowell RM. Opportunities for lignocellulosic materials and composites. In: Rowell RM, Schultz TP, Narayan R., Ed. Emerging technologies for materials and chemicals from biomass. American Chemical Society, 1992. p. 2–12.
[9] Naik S, Goud V V, Rout PK, Jacobson K, Dalai AK. Characterization of Canadian biomass for alternative renewable biofuel. Renew Energy 2010, 35, 1624–31.
[10] Ilyas U, Ahmed S, Majeed A, Nadeem M. Exploration of indigenous agrowastes for cellulase production by *Aspergillus niger*. African J Biotechnol 2012, 11, 9276–9.
[11] Howard RL, Abotsi E, Jansen van Rensburg EL, and Howard S. Lignocellulose biotechnology: issues of bioconversion and enzyme production. African J Biotechnol 2003, 2, 602–19.
[12] Prasad S, Singh A, Joshi HC. Ethanol as an alternative fuel from agricultural, industrial and urban residues. Resour Conserv Recycl 2007, 50, 1–39.
[13] Li H, Kim N-J, Jiang M, Kang JW, Chang HN. Simultaneous saccharification and fermentation of lignocellulosic residues pretreated with phosphoric acid-acetone for bioethanol production. Bioresour Technol 2009, 100, 3245–51.
[14] Raveendran K, Ganesh A, Khilar KC. Influence of mineral matter on biomass pyrolysis characteristics. Fuel 1995, 74, 1812–22.
[15] Okeke BC, Obi SKC. Lignocellulose and sugar compositions of some agro-waste materials. Bioresour Technol 1994, 47, 283–4.
[16] Kim TH, Kim JS, Sunwoo C, Lee YY. Pretreatment of corn stover by aqueous ammonia. Bioresour Technol 2003, 90, 39–47.
[17] Merino S, Cherry J. Progress and challenges in enzyme development for biomass utilization. Adv Biochem Eng Biotechnol 2007, 108, 95–120.
[18] Malherbe S, Cloete TE. Lignocellulose biodegradation: Fundamentals and applications. Rev Environ Sci Biotechnol 2002, 1, 105–14.
[19] Nanda S, Mohanty P, Pant K, Naik S, Kozinski J, Dalai A. Characterization of North American lignocellulosic biomass and biochars in terms of their candidacy for alternate renewable fuels. BioEnergy Res 2013;6(2):663–77.
[20] Ballesteros M, Oliva JM, Negro MJ, Manzanares P, Ballesteros I. Ethanol from lignocellulosic materials by a simultaneous saccharification and fermentation process (SFS) with Kluyveromyces marxianus CECT 10875. Process Biochem 2004, 39, 1843–8.
[21] Tang J, Chen K, Huang F, xu J, Li J. Characterization of the pretreatment liquor of biomass from the perennial grass, *Eulaliopsis binata*, for the production of dissolving pulp. Bioresour Technol 2013, 129, 548–52.
[22] Lee J. Biological conversion of lignocellulosic biomass to ethanol. J Biotechnol 1997, 56, 1–24.
[23] Ferreira-Leitao V, Perrone C, Rodrigues J, Franke A, Macrelli S, Zacchi G. An approach to the utilisation of $CO_2$ as impregnating agent in steam pretreatment of sugar cane bagasse and leaves for ethanol production. Biotechnol Biofuels 2010, 3, 1–8.
[24] Mansoer Z, Reeves DW, Wood CW. Suitability of Sunn Hemp as an alternative late-summer legume cover crop. Soil Sci Soc Am J 1997, 61, 246–53.

[25] Kabel MA, van der Maarel MJ, Klip G, Voragen AG, Schols HA. Standard assays do not predict the efficiency of commercial cellulase preparations towards plant materials. Biotechnol Bioeng 2006, 93, 56–63.
[26] Bridgeman TG, Jones JM, Shield I, Williams PT. Torrefaction of reed canary grass, wheat straw and willow to enhance solid fuel qualities and combustion properties. Fuel 2008, 87, 844–56.
[27] Agbor VB, Cicek N, Sparling R, Berlin A, Levin DB. Biomass pretreatment: Fundamentals toward application. Biotechnol Adv 2011, 29, 675–85.
[28] Chandra R, Bura R, Mabee W, Berlin A, Pan X, Saddler J. Substrate pretreatment: the key to effective enzymatic hydrolysis of lignocellulosics? Adv Biochem Eng Biotechnol 2007, 108, 67–93.
[29] Jeoh T, Ishizawa CI, Davis MF, Himmel ME, Adney WS, Johnson DK. Cellulase digestibility of pretreated biomass is limited by cellulose accessibility. Biotechnol Bioeng 2007, 98, 112–22.
[30] Brodeur G, Yau E, Badal K, Collier J, Ramachandran KB, Ramakrishnan S. Chemical and Physicochemical Pretreatment of Lignocellulosic Biomass: A Review. Enzyme Res2011, Article ID-787532, 1–17.
[31] Verardi A, Bari I De, Ricca E, Calabrò V. Hydrolysis of lignocellulosic biomass: current status of processes and technologies and future perspectives. In: Lima MAP, editor. Bioethanol. In Tech, 2012. p. 95–122.
[32] Zhu S, Wang K, Huang W, et al. Acid-catalyzed hydrolysis of lignocellulosic biomass in ionic liquids for ethanol production: Opportunities & challenges. BioResources 2016, 11, 3–5.
[33] Timung R, Naik Deshavath N, Goud V V., Dasu V V. Effect of subsequent dilute acid and enzymatic hydrolysis on reducing sugar production from sugarcane bagasse and spent Citronella biomass. J Energy 2016, 2016, 1–12.
[34] Radeva G, Valchev I, Petrin S, Valcheva E, Tsekova P. Kinetic study of the enzyme conversion of steam exploded Paulownia tomentosa to glucose. BioResources 2012, 7, 412–21.
[35] Leu S-Y, Zhu JY. Substrate-related factors affecting enzymatic saccharification of lignocelluloses: Our recent understanding. BioEnergy Res 2013, 6, 405–15.
[36] Van Dyk JS, Pletschke BI. A review of lignocellulose bioconversion using enzymatic hydrolysis and synergistic cooperation between enzymes-Factors affecting enzymes, conversion and synergy. Biotechnol Adv 2012, 30, 1458–80.
[37] Menon V, Rao M. Trends in bioconversion of lignocellulose: Biofuels, platform chemicals & biorefinery concept. Prog Energy Combust Sci 2012, 38, 522–50.
[38] Paramjeet S, Manasa P, Korrapati N. Biofuels: Production of fungal-mediated ligninolytic enzymes and the modes of bioprocesses utilizing agro-based residues. Biocatal Agric Biotechnol 2018, 14, 57–71.
[39] Barros-Rios J, Romani A, Peleteiro S, Garrote G, Ordas B. Second-generation bioethanol of hydrothermally pretreated stover biomass from maize genotypes. Biomass and Bioenergy 2016, 90, 42–9.
[40] Patel H, Chapla D, Shah A. Bioconversion of pretreated sugarcane bagasse using enzymatic and acid followed by enzymatic hydrolysis approaches for bioethanol production. Renew Energy,2017, 109, 323–31.
[41] von Sivers M, Zacchi G. Ethanol from lignocellulosics: A review of the economy. Bioresour Technol, 1996, 56, 131–40.
[42] Ishizaki H, Hasumi K. Ethanol Production from Biomass. In: Tojo S, Hirasawa T, editors. Research Approaches to Sustainable Biomass Systems. Academic Press; 2014. p. 243–58.
[43] Gallego LJ, Escobar A, Peñuela M, Peña JD, Rios LA. King Grass: A promising material for the production of second-generation butanol. Fuel 2015, 143, 399–403.
[44] Joshi SM, Waghmare JS, Sonawane KD, Waghmare SR. Bio-ethanol and bio-butanol production from orange peel waste. Biofuels 2015, 6, 55–61.

[45] Maiti S, Gallastegui G, Sarma SJ, et al. A re-look at the biochemical strategies to enhance butanol production. Biomass and Bioenergy 2016, 94, 187–200.
[46] Chandra R, Takeuchi H, Hasegawa T. Hydrothermal pretreatment of rice straw biomass: A potential and promising method for enhanced methane production. Appl Energy 2012, 94, 129–40.
[47] Asadi N, Zilouei H. Optimization of organosolv pretreatment of rice straw for enhanced biohydrogen production using *Enterobacter aerogenes*. Bioresour Technol 2017, 227, 335–44.
[48] Kumar G, Sivagurunathan P, Sen B, et al. Research and development perspectives of lignocellulose-based biohydrogen production. Int Biodeterior Biodegradation 2017, 119, 225–38.
[49] Gonzales RR, Kumar G, Sivagurunathan P, Kim SH. Enhancement of hydrogen production by optimization of pH adjustment and separation conditions following dilute acid pretreatment of lignocellulosic biomass. Int J Hydrogen Energy 2017, 42, 27502–11.
[50] Mesa L, Martínez Y, Barrio E, González E. Desirability function for optimization of dilute acid pretreatment of sugarcane straw for ethanol production and preliminary economic analysis based in three fermentation configurations. Appl Energy 2017, 198, 299–311.
[51] Domínguez E, Romaní A, Domingues L, Garrote G. Evaluation of strategies for second generation bioethanol production from fast growing biomass Paulownia within a biorefinery scheme. Appl Energy 2017, 187, 777–89.
[52] Martínez-Patiño JC, Ruiz E, Romero I, Cara C, López-Linares JC, Castro E. Combined acid/ alkaline-peroxide pretreatment of olive tree biomass for bioethanol production. Bioresour Technol 2017, 239, 326–35.
[53] Tomás-Pejó E, Fermoso J, Herrador E, et al. Valorization of steam-exploded wheat straw through a biorefinery approach: Bioethanol and bio-oil co-production. Fuel 2017, 199, 403–12.
[54] Chandel AK, Singh O V, Venkateswar Rao L, Chandrasekhar G, Lakshmi Narasu M. Bioconversion of novel substrate Saccharum spontaneum, a weedy material, into ethanol by *Pichia stipitis* NCIM3498. Bioresour Technol 2011, 102, 1709–14.
[55] Das A, Paul T, Jana A, et al. Bioconversion of rice straw to sugar using multizyme complex of fungal origin and subsequent production of bioethanol by mixed fermentation of *Saccharomyces cerevisiae* MTCC 173 and *Zymomonas mobilis* MTCC 2428. Ind Crops Prod 2013, 46, 217–25.
[56] Qureshi N, Saha BC, Cotta MA. Butanol production from wheat straw by simultaneous saccharification and fermentation using *Clostridium beijerinckii*: Part II – Fed-batch fermentation. Biomass and Bioenergy 2008, 32, 176–83.
[57] Qureshi N, Saha BC, Hector RE, et al. Production of butanol (a biofuel) from agricultural residues: Part II – Use of corn stover and switchgrass hydrolysates. Biomass and Bioenergy 2010, 34, 566–71.
[58] Gao K, Rehmann L. ABE fermentation from enzymatic hydrolysate of NaOH-pretreated corncobs. Biomass and Bioenergy 2014, 66, 110–5.
[59] Moradi F, Amiri H, Soleimanian-Zad S, Ehsani MR, Karimi K. Improvement of acetone, butanol and ethanol production from rice straw by acid and alkaline pretreatments. Fuel 2013, 112, 8–13.
[60] Zhang Y, Chen X, Gu Y, Zhou X. A physicochemical method for increasing methane production from rice straw : Extrusion combined with alkali pretreatment. 2015, 160, 39–48.
[61] Khor CW, Rabaey K, Vervaeren H. Low temperature calcium hydroxide treatment enhances anaerobic methane production from (extruded) biomass. Bioresour Technol 2015, 176, 181–8.
[62] Arreola-vargas J, Ojeda-castillo V, Snell-castro R, Corona-gonzález RI, Alatriste mondragón F, Méndez-acosta HO. Methane production from acid hydrolysates of *Agave tequilana* bagasse : Evaluation of hydrolysis conditions and methane yield. Bioresour Technol 2015, 181, 191–9.

[63] Xihui K, Yongming S, Lianhua L, Xiaoying K, Zhenhong Y. Improving methane production from anaerobic digestion of *Pennisetum hybrid* by alkaline pretreatment. Bioresour Technol 2018, 255, 205–12.

[64] Baba Y, Matsuki Y, Mori Y, et al. Pretreatment of lignocellulosic biomass by cattle rumen fluid for methane production : Bacterial flora and enzyme activity analysis. J Biosci Bioeng 2017, 123(4), 489–96.

[65] El-Mashad HM. Biomethane and ethanol production potential of *Spirulina platensis* algae and enzymatically saccharified switchgrass. Biochem Eng J,2015, 93, 119–27.

[66] Panagiotopoulos IA, Bakker RR, de Vrije T, Koukios EG, Claassen PAM. Pretreatment of sweet sorghum bagasse for hydrogen production by *Caldicellulosiruptor saccharolyticus*. Int J Hydrogen Energy 2010, 35, 7738–47.

[67] Lopez-Hidalgo AM, Sánchez A, De León-Rodríguez A. Simultaneous production of bioethanol and biohydrogen by *Escherichia coli* WDHL using wheat straw hydrolysate as substrate. Fuel 2017, 188, 19–27.

[68] Cheng X-Y, Liu C-Z. Fungal pretreatment enhances hydrogen production via thermophilic fermentation of cornstalk. Appl Energy 2012, 91, 1–6.

[69] Pan C, Fan Y, Hou H. Fermentative production of hydrogen from wheat bran by mixed anaerobic cultures. Ind Eng Chem Res 2008, 47, 5812–8.

[70] Lai Z, Zhu M, Yang X, Wang J, Li S. Optimization of key factors affecting hydrogen production from sugarcane bagasse by a thermophilic anaerobic pure culture. Biotechnol Biofuels 2014, 7, 119.

[71] Monlau F, Barakat A, Steyer JP, Carrere H. Comparison of seven types of thermo-chemical pretreatments on the structural features and anaerobic digestion of sunflower stalks. Bioresour Technol 2012, 120, 241–7.

[72] Ibrahim CO. Development of applications of industrial enzymes from Malaysian indigenous microbial sources. Bioresour Technol 2008, 99, 4572–82.

[73] Ravindran R, Jaiswal AK. Microbial enzyme production using lignocellulosic food industry wastes as feedstock: A review. Bioengin 2016, 3, 1–22.

[74] Gustavo FNE, Cesar BA, Shuri I, et al. Artificial intelligence approach for high level production of amylase using *Rhizopus microsporus var. oligosporus* and different agro-industrial wastes. J Chem Technol Biotechnol 2016, 92(3), 684–92.

[75] Uygut MA, Tanyildizi MŞ. Determination of effective parameters for alpha-amylase production in a modified rotating drum bioreactor. Arab J Sci Eng 2018, 43, 3381–91.

[76] Escaramboni B, Fernández Núñez EG, Carvalho AFA, de Oliva Neto P. Ethanol biosynthesis by fast hydrolysis of cassava bagasse using fungal amylases produced in optimized conditions. Ind Crops Prod 2018, 112, 368–77.

[77] Singh S, Singh S, Bali V, Sharma L, Mangla J. Production of fungal amylases using cheap, readily available agriresidues, for potential application in textile industry. Biomed Res Int 2014;Article ID 215748, 1–9.

[78] Novelli PK, Barros MM, Fleuri LF. Novel inexpensive fungi proteases: Production by solid state fermentation and characterization. Food Chem 2016, 198, 119–24.

[79] Kumar A, Dutt D, Gautam A. Production of crude enzyme from *Aspergillus nidulans* AKB-25 using black gram residue as the substrate and its industrial applications. J Genet Eng Biotechnol 2016, 14, 107–118.

[80] Gautam, A, Kumar A, Dutt, D. Production and characterization of cellulase-free xylanase by *Aspergillus flavus* ARC-12 and its application in pre-bleaching of ethanol-soda pulp of *Eulaliopsis binata*. Res J Biotechnol 2017, 12, 63–71.

[81] Pandya JJ, Gupte A. Production of xylanase under solid-state fermentation by *Aspergillus tubingensis* JP-1 and its application. Bioprocess Biosyst Eng 2012, 35, 769–79.

[82] Gautam A, Kumar A, Bharti AK, Dutt D. Rice straw fermentation by *Schizophyllum commune* ARC-11 to produce high level of xylanase for its application in pre-bleaching. J Genet Eng Biotechnol 2018, https://doi.org/10.1016/j.jgeb.2018.02.006
[83] Pathak P, Bhardwaj N, Singh A. Production of crude cellulase and xylanase from *Trichoderma harzianum* PPDDN10 NFCCI-2925 and Its application in photocopier waste paper recycling. Appl Biochem Biotechnol 2014, 172, 3776–97.
[84] Bharti AK, Kumar A, Kumar A, Dutt D. Wheat bran fermentation for the production of cellulase and xylanase by *Aspergillus niger* NFCCI 4113. Res J Biotechnol 2018, 13, 11–18.
[85] Silva VCT, Coto ALS, Souza RC, Neves MBS, Gomes E, Bonilla-Rodriguez GO. Effect of pH, temperature, and chemicals on the endoglucanases and beta-glucosidases from the thermophilic fungus *Myceliophthora heterothallica* F.2.1.4. obtained by solid-state and submerged cultivation. Biochem Res Int 2016, Article ID 9781216, 1–9.
[86] Bharti AK, Kumar A, Kumar A, Dutt D. Exploitation of Parthenium hysterophorous biomass as low-cost substrate for cellulase and xylanase production under solid-state fermentation using *Talaromyces stipitatus* MTCC 12687. J Radiat Res Appl Sci.https://doi.org/10.1016/j.jrras.2018.01.003
[87] Kumar A, Gautam A, Dutt D. Co-cultivation of *Penicillium* sp. AKB-24 and *Aspergillus nidulans* AKB-25 as a cost-effective method to produce cellulases for the hydrolysis of pearl millet stover. Fermentation 2016, 2, 1–22.
[88] Wang Q, Chen L, Yu D, Lin H, Shen Q, Zhao Y. Excellent waste biomass- degrading performance of Trichoderma asperellum T-1 during submerged fermentation. Sci Total Environ 2017, 609, 1329–39.
[89] Bagewadi Z, Mulla S, Ninnekar H. Optimization of endoglucanase production from *Trichoderma harzianum* strain HZN11 by central composite design under response surface methodology. Biomass Conv Bioref 2017, 8, 305–316.
[90] Postemsky PD, Bidegain MA, González-Matute R, Figlas ND, Cubitto MA. Pilot-scale bioconversion of rice and sunflower agro-residues into medicinal mushrooms and laccase enzymes through solid-state fermentation with *Ganoderma lucidum*. Bioresour Technol 2017, 231, 85–93.
[91] Thiribhuvanamala G., Kalaiselvi G, Parthasarathy S, Anusha B. Induction of lignolytic enzyme activities in different agro residues by the white rot fungi, *Pleurotus sajar-caju*. Int J Chem Stud 2017, 5, 89–94.
[92] Wang Z, Liu J, Ning Y, Liao X, Jia Y. Eichhornia crassipes: Agro-waster for a novel thermostable laccase production by *Pycnoporus sanguineus* SYBC-L1. J Biosci Bioeng 2017, 123, 163–9.
[93] Sauer M, Porro D, Mattanovich D, Branduardi P. Microbial production of organic acids: expanding the markets. Trends Biotechnol 2008, 26, 100–8.
[94] Alonso S, Rendueles M, Díaz M. Microbial production of specialty organic acids from renewable and waste materials. Crit Rev Biotechnol 2015, 35, 497–513.
[95] Khan I, Qayyum S, Maqbool F, Hayat A, Sharif M. Microbial organic acids production, biosynthetic mechanism and applications-Mini review. Indian J Geo Mar Sci 2017, 46, 2165–74.
[96] Crolla A, Kennedy KJ. Optimization of citric acid production from *Candida lipolytica* Y-1095 using n-paraffin. J Biotechnol 2001, 89, 27–40.
[97] Xu Q, Li S, Huang H, Wen J. Key technologies for the industrial production of fumaric acid by fermentation. Biotechnol Adv [Internet] 2012;30(6):1685–96.
[98] Woiciechowski AL, Soccol CR, Ramos LP, Affonso LF. Screening of several Rhizopus strains to produce fumaric acid by biological conversion of hemicellulosic hydrolysates obtained by steam explosion. V Simpósio de Hidrólise Enzimática de Biomassa, 2001, Maringá, PR. Proceedings (CD rom edition). Maringá, PR: Universidade Estadual de Maringá; 1996.

[99] Alcantara J, Mondala A, Hughey L, Shields S. Direct Succinic Acid Production from Minimally Pretreated Biomass Using Sequential Solid-State and Slurry Fermentation with Mixed Fungal Cultures. Ferment.. 2017;3(3).
[100] Enrique C-C, Cristina G-F, Mercedes B, Elia T-P. Biotechnological advances in lactic acid production by lactic acid bacteria: lignocellulose as novel substrate. Biofuels, Bioprod Biorefining 2018, 12(2), 290–303.
[101] Zheng J, Gao M, Wang Q, et al. Enhancement of l-lactic acid production via synergism in open co-fermentation of Sophora flavescens residues and food waste. Bioresour Technol 2017, 225, 159–64.
[102] Donot F, Fontana A, Baccou JC, Schorr-Galindo S. Microbial exopolysaccharides: Main examples of synthesis, excretion, genetics and extraction. Carbohydr Polym 2012, 87, 951–62.
[103] Leathers TD, Nunnally MS, Stanley AM, Rich JO. Utilization of corn fiber for production of schizophyllan. Biomass and Bioenergy 2016, 95, 132–6.
[104] An C, Ma S, Chang F, Xue W. Efficient production of pullulan by *Aureobasidium pullulans* grown on mixtures of potato starch hydrolysate and sucrose. Brazilian J Microbiol 2017, 48, 180–5.
[105] Yang J, Zhang Y, Zhao S, Zhou Q, Xin X, Chen L. Statistical optimization of medium for pullulan production by *Aureobasidium pullulans* NCPS2016 using fructose and soybean meal hydrolysates. Mol. 2018, 23, 1–16.
[106] López-Linares JC, Romero I, Cara C, Castro E, Mussatto SI. Xylitol production by *Debaryomyces hansenii* and *Candida guilliermondii* from rapeseed straw hemicellulosic hydrolysate. Bioresour Technol 2018, 247, 736–43.
[107] Kumar V, Krishania M, Preet Sandhu P, Ahluwalia V, Gnansounou E, Sangwan RS. Efficient detoxification of corn cob hydrolysate with ion-exchange resins for enhanced xylitol production by *Candida tropicalis* MTCC 6192. Bioresour Technol 2018, 251, 416–9.
[108] Albuquerque TL de, da Silva IJ, de Macedo GR, Rocha MVP. Biotechnological production of xylitol from lignocellulosic wastes: A review. Process Biochem 2014, 49(11), 1779–89.

Ashish Kumar Singh and Neelam Verma

# 4 Plants and plant-derived materials used for biosensor development

**Abstract:** Biosensor, which is used for the detection of particular analytes, is an analytical device that works based on physicochemical reaction. Biosensors are promising bio-devices, which can be used as substitutes in traditional analytical methods, for rapid, easy, economical and reliable analysis. This chapter discusses the applications and development of biosensors based on plants and plant-derived materials as biological elements.

**Keywords:** Biosensor, plant, immobilization, transducers

## 4.1 Introduction

The rapid, sensitive and selective detection of desired materials is important in several fields like food and bioprocessing technology, clinical analysis and environmental monitoring [1–5]. Research into the development and further growth of biosensors has increased greatly. Unlike conventional analytical methods, biosensors lead to rapid, easy, economic, sensitive and selective methods to monitor and quantify the choice of analytes in real samples. In essence, a biosensor couples a recognition event or a chemical reaction by a biological entity with a transducer, which then converts this chemical reaction into a readable signal. The biological materials are immobilized onto the transducer and specifically recognize the target molecule, and the transducer system reads the chemical reaction and converts it into quantifiable data or signal. This type of combinatorial work provides an analytical method that is comparable to the already-established and traditional analytical methods [6].

Biosensors are classified according to the transducer used or the transducing system. Although various transducers are used for biosensor-based detection methods, the most frequently used transducers are as follows:

- electrochemical transducers, e.g. amperometric, conductometric and potentiometric;
- optical transducers, such as absorbance-based colorimetric (UV–vis), emission-based bioluminescence (luminometers), and fluorescence (fiber-optic biosensors);
- heat-sensitive transducers, such as thermistor-based biosensors; and
- mass-sensitive transducers, such as piezoelectric, e.g. quartz crystal microbalance (QCM).

https://doi.org/10.1515/9783110563337-004

Biosensors can also be classified according to the biological materials used or immobilized as follows:
- enzymes, which recognize particular substrates and catalyze specific chemical reactions;
- antibodies, which specifically bind with the respective antigens;
- oligonucleotides, which undergo base pairing with complementary sequences;
- whole cells, where target analyte is a metabolites that utilized by the cells;
- tissues, mostly used as the source of enzyme; and
- biomimetic materials, which are also known as synthetic bio-receptors.

Each of these biosensor techniques are based on either different transducers or biological components, and hence have some advantages and disadvantages. Use of electrochemical-based approaches for the development of biosensors is very promising due to its simple fabrication, high sensitivity, less environmental impacts, onsite usability, and ease of miniaturization and economic fabrication. These biosensors have several advantages: they have excellent operational stability, long storage, relatively short response times and high sensitivity. In addition, other interferences (e.g. ammonia) do not have any impact on them, and nearly all biosensors achieve the required analytical range. The disadvantages of these biosensors are there are chances of losing enzyme activity during immobilization and there are increased chances of ionic interference. Optical detection is usually based on the measurement of luminescent, fluorescent, colorimetric or other optical signals produced by the interaction of the bio-component with the analytes, and it correlates the observed optical signal with the concentration of the target compounds. Optical sensing techniques are especially attractive in high-throughput screening since they enable biosensors to monitor multiple analytes simultaneously [7]. In such biosensors, optical fibers are used to guide the light waves to suitable detectors like an electrode or a semiconductor [8, 9]. An advantage of optical sensors is their ability to resolve information from different analytes via their discrete wave bands. This would allow one to couple an ammonia sensor with a sensor for a different analyte in the same membrane, but collect the readout information at separate wavelengths. However, they suffer from disadvantages like laborious labeling processes, which may also interfere with the function of a biomolecule. Quantitative analysis is challenging due to the fluorescence signal bias as the number of fluorophores on each molecule cannot be precisely controlled. Piezoelectric biosensors operate on the principle of coating the surface of the biosensor with selectively binding biologically active substances. The piezoelectric transducer allows a binding event to be converted into a measurable, for instance, a resonance frequency change. Piezoelectric biosensors offer a real-time output, the simplicity of a wider working pH range and cost effectiveness. Possible disadvantages of the biosensors include the lack of sensitivity as well as interferences from the liquid media where the analysis takes place [2]. Thermal detection biosensors exploit one of the fundamental properties of biological reactions, namely absorption or production of heat, which in turn changes the temperature of the medium in which the reaction takes place. These biosensors are developed by combining immobilized enzyme molecules with temperature sensors. When the analyte comes in contact with the enzyme, the heat reaction of the enzyme is measured and calibrated against the analyte concentration. The total heat produced or absorbed is proportional to the molar enthalpy and the total number of molecules in the reaction. The measurement of temperature is typically accomplished via a thermistor, and such devices are known as enzyme thermistors. Their high sensitivity to thermal changes makes thermistors ideal for such applications. Unlike other transducers, thermal biosensors do not need frequent recalibration and are insensitive to the optical and electrochemical properties of the sample. Common applications of this type of biosensor include clinical monitoring, process monitoring and environmental monitoring [10, 11].

This chapter focusses on the development of biosensors that exploit plant-derived materials as bio-components, among them plant-tissue-based biosensors being the most reported.

## 4.2 Plant-tissue-based biosensors

In the early 1980s, another progress within the development of tissue-based electrodes was the invention that materials of plant origin can be often used as effective biocatalysts. In 1981, Kuriyama and Rechnitz created the primary plant-tissue-based biosensor, utilizing a tissue slice from a yellow squash and a carbon dioxide electrode [12]. The enzyme activity of this conductor arose from the enzyme (glutamate decarboxylase) present within the tissue of yellow squash. The enzyme glutamate decarboxylase catalyzes the conversion of glutaminic acid to carbon dioxide that is detected by the carbon dioxide electrode and produces a possible response associated with the concentration of glutaminic acid within the sample. This method shows the primary use of intact plant materials as a bio-component for the fabrication of the biosensor. Several plant-tissue-based biosensors have since been developed exploiting tissue materials from growing parts of plants (e.g. young leaves, blossoms) or their nutrient storage systems (e.g. vegetables, fruits) [13].

### 4.2.1 Importance of biosensors based on plant tissues

There are numerous advantages of biosensors based on plant tissues:
- highly stable and show high enzyme activity in their optimum conditions;
- easily available and very economical for the development of biosensors;
- high reliability;

- increased diverse types of plants and plant-derived materials lead to a wider range of biosensor constructions;
- long storage stability of biosensors because of the long stability of biological materials; and
- elimination of dreary and time-consuming enzyme extraction and purification steps, with the resultant reduction in cost of biosensor development and applications.

However, they also have certain drawbacks. They show poor specificity and have longtime intervals because of the presence of protein in the intact tissue, hindering the flow. The usual method used in the development of biosensors is either supporting a skinny slice of a plant part onto the conductor's surface or entrapping it into carbon paste. The biosensor-supported membrane stops the escape of catalysts from the conductor surface but modifies the substrate flow into it. This flow obstruction prolongs the biosensor's time interval. However, electrodes based on carbon paste have been found to reduce the response time in seconds. Usually, biosensors based on plant tissues are chemical transducers, either amperometric or potentiometric. However, optical transducers, like visible radiation or luminescence, can provide faster response times along with higher sensitivities. Because of the large diversity of plants and its derived materials, various biosensors are being developed. Next, a number of plant-tissue-based biosensors are discussed.

### 4.2.2 Plant tissue as a source of polyphenol oxidase

Banana pulp slices are largely utilized in the development of biosensors due to the high content of the enzyme polyphenol oxidase (PPO), which recognizes various phenolic compounds. Sidwell and Rechnitz first used banana tissue as a source of PPO for biosensor development [13], and used the term "bananatrode". The biosensor was developed using a thin slice of banana pulp tissue onto the surface of a Clark-type oxygen electrode to determine the oxygen consumption required during the conversion of dopamine into 1,2-benzoquinone. Later, Wang and Lin [14] used banana pulp tissue as a source of PPO and immobilized this enzyme into the carbon paste matrix, thereby reducing the reaction time of dopamine. Rather than measure the oxygen utilization, they quantified the reduction of the compound's spinoff back to the catechol at the negative potentials. Furthermore, Eggins *et al.* [15] also used the same banana pulp tissue as a working electrode to measure catechins with dimeric and trimeric forms, which are used as indicators for the quality check of beer production. However, their results were inadequate most likely because of the flavonoids already found in banana. Similarly, wet or dried apple and potato tissues, which also have PPO, can also be used to achieve good analytical data such as those obtained by typical colorimetrical and natural techniques. Cummings *et al.* [16] used apple tissues entrapped in the matrix of carbon paste and further characterized the electrochemical, analytical and morphological behaviors of electrodes to optimize the performance of the developed biosensor. They concluded that higher ratios of plant tissues to the matrix of carbon paste produce non-homogeneous surfaces that result in increased water penetration, better range of accessible chemical action sites inside the network and improved PPO accessibility.

Apart from tissues of bananas, potatoes and apples, other plants have also been exploited due to their high PPO contents. A detection system based on flow injection analysis was developed to analyze catechol using coconut tissue (*Cocus nucifera*) as a bio-catalyst [17]. However, the tissue was not immobilized on the surface of the working electrode but entrapped in the capillary of the glass reactor. Since the bio-component was not in close proximity with the transducer element, the device cannot be called a true "biosensor". The applied potential for catechol detection was found to be–0.10 V vs. Ag/AgCl with the detection limit of 2 μM and the rate of detection being up to 60–90samples/h. Despite the entangling of tissue by sorption of synthetic resin polymers fashioned as other metabolites using the catalyst method, the system can be run for several days with only a small loss of enzyme activity. The system proved very advantageous for the detection of catechol content in stream water and sedge waste material.

Avocado (*Persea americana*) tissue is also used as a great source of PPO. An amperometric biosensor has been developed for paracetamol detection using a mixture of graphite/Vaseline with avocado tissues as the source of enzyme PPO [18]. High dose of paracetamol could cause liver damage or failure. PPO in avocado tissues catalyzed the paracetamol oxidization to N-acetyl-p-benzoquinone, which was further reduced and regenerated to paracetamol and detected at the potential of −0.12 V vs. Ag/AgCl. The detection limit of the developed biosensor was found to be 88 μM,which was in good agreement with the results obtained by the standard procedure.

In the pharma industries, biosensors based on fibers obtained from the fruits of palm tree are used to detect the hormone epinephrine [19]. Here, PPO catalyzed the oxidation of epinephrine to epinephrine quinone, and it was electrochemically reduced at the potential of −0.10 V vs. Ag/AgCl. The enzyme was immobilized in the carbon paste matrix for the fabrication of the biosensor and also assembled with the flow injection analysis system for continuous detection. The detection limit of the developed biosensor was found to be 15 μM. The developed biosensor was also validated by comparing its results with the results obtained from the spectrophotometer, and very good agreement was found.

Some of the major disadvantages of biosensors developed with the use of biocatalysts other than those based on plants are the very short shelf-life and loss of response to target. The long-term storage stability of biocatalysts from plant tissues can address these above-mentioned problems. Here, a biosensor has been developed based on potato tissue (*Solanum tuberosum*) for the detection of phenols and phenolic compounds. The detection principle is based on the inhibition of enzyme PPO by herbicide atrazine in water samples [20]. The analytical performance of this biosensor in terms of the detection limit was found to be 10 μM;moreover, there was negligible response to interference molecules like organophosphorus or salt pesticides including paraoxon, malathion, insect powder and aldicarb. The storage stability of the developed biosensor was found to be 8–14 days, and although it was considerably shorter than the ones used for catechol or phenol determination, this can be improved by washing its probe after every measurement to remove any traces of the samples. In any case, the utilization of tissue rather than pure protein made it easy to replace the biocatalyst layer, leading to an inexpensive "partially disposable" biosensor.

## 4.2.3 Plant tissue as a source of peroxidase for biosensor development

One of the most useful enzymes for biosensor development is peroxidase (POD). Because of its occurrence in several plants, a range of biosensors are being delineated. Wang *et al.* [21] used horseradish tissue as a potential source of POD enzyme and immobilized it onto the surface of a carbon disk electrode. They developed a biosensor for the detection of 2-butanone peroxide with a response time of only 20 s. The significance of this method is that the developed biosensor successfully works in an organic solvent medium. Vieira & Fatibello-Filho [22] also fabricated a biosensor that can be used in organic solvent media based on the same enzyme POD. They slightly modified the source of the enzyme and used a sweet potato tissue (*Ipomea batatas*) mixture with paraffin/graphite, which was used to detect hydroquinone in cosmetic materials. The occurrence of oxide, it is a natural substrate in the enzyme POD where it oxidizes hydroquinone to the para-quinone, which is reduced back to hydroquinone at the potential of −0.22 V vs. Ag/AgCl (Figure 4.1). The detection limit was found to be about 8 μM, equivalent to that obtained by the standard method. The biosensor approach allowed the direct detection of hydroquinone in cosmetic creams after solubilization, while not hydroquinone extraction to the aqueous.

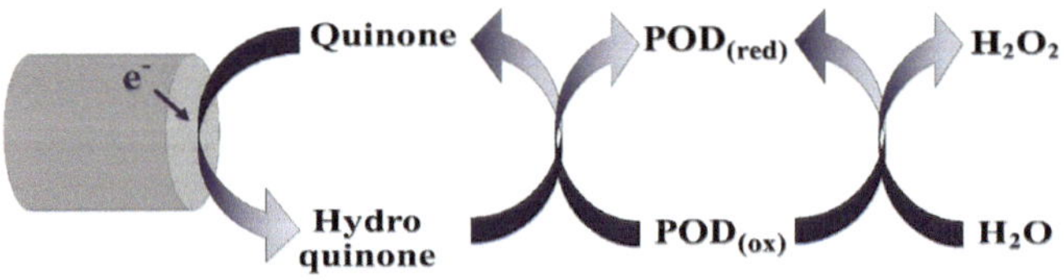


Figure 4.1: Working principle of polyoxidase (sweet potato tissue) based biosensor for polyphenol.

The crude extract from gilo (*Solanum gilo*) was immobilized in chitosan biopolymer to develop a biosensor to detect hydroquinone [23] and rutin [24]. Gilo contains high titer of POD; however, this enzyme is not commercially produced from this plant due to the high cost of the purification and extraction procedure. In chitosan, the amine group composite obtained with the base-forming deacetylation of chitin is extremely reactive and forms strong interactions with POD present in the tissue of gilo plants. The developed biosensors showed high sensitivity (detection limit of 2 μM for hydroquinone and 20nM for rutin), were consistent (1% RSD) and were found to be extremely stable. They also coupled the tissue from green bean (*Phaseolus vulgaris*) to chitin matrix to develop a biosensor for the detection of caffeic acid in wine [25].This biosensor too was highly sensitive (detection limit of 2 μM), consistent (2.2% RSD) and stable (10 months).

Tissues from the asparagus plant are also a good source of POD. Oungpipat *et al.* [26] co-immobilized ground tissue of asparagus (*Asparagus officinalis*) and ferrocene in a carbon paste matrix to develop a glass capillary-based oxide biosensor. The ferrocene acts as a redox mediator, which facilitates electron transfer between the biocatalyst and the electrode. It resulted in decreased operating potential, and therefore in reduced direct reaction of the interference molecules (Figure 4.2). The developed biosensor had a detection limit of 0.4 μM and a reaction time of 2s. Another similar biosensor with slight modifications has been developed and applied to determine fluoride, which is beneficial for teeth and bones but harmful beyond a specific (1.5 mg/L) concentration [27]. During the design process, the detection limit of fluoride on POD was determined. The detection limit of the biosensor was found to be 0.5 mg/L, indicating its relevance in real samples. The industrial tablet formulations were analyzed, and the results were found to be similar to those obtained with a fluoride selective biosensor.

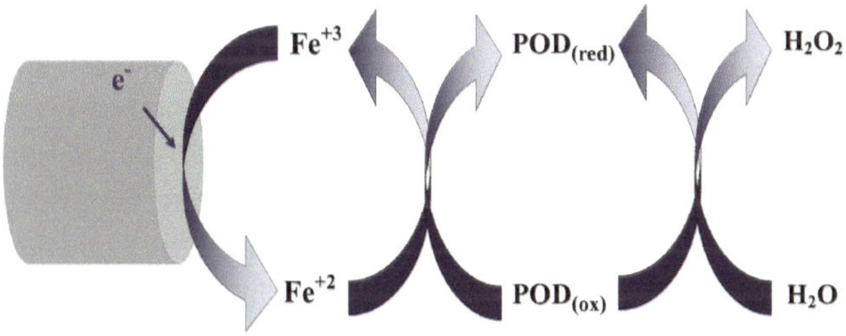


Figure 4.2: Working principle of ferrocene-mediated asparagus tissue-based biosensor.

Kozan *et al.* [28] reported the use of coconut fibers to fabricate a biosensor for the detection of peroxide. The developed biosensor operated at the potential of −0.15 V vs. Ag/AgCl, responded speedily and its detection limit was found to be 40 μM for peroxide. Moreover, 50 measurements were performed with the same electrode and no significant decrease in the sensor response was found; moreover, the storage time period was 90 days. The biosensor was tested with medicinal items, like antiseptic dental rinse solution, solution for contact lenses cleaning and hair coloring gel, and the obtained results were found to correlate with those obtained by the standard detection technique.

Most of the biosensors have been developed by the commercially available plant enzymes and other plant-derived materials. The freshness, maturity level, climate or seasonal conditions affect the reproducibility of the results. To overcome these drawbacks, Navaratne and Rechnitz [29] explored tissues from *in vitro*-cultured tobacco (*Nicotiana tabacum*) callus. These tissues served as a potential source of POD along with high and reproducible catalyst activity. The *in vitro* culture involves the use of explants from the suitable plant body followed by their aseptic inoculation in a specific growth medium. The growth media composition, culture conditions and other factors are reproducible, and therefore, uniform tissues can be obtained. The results obtained from the biosensors (based on different biocatalyst sources) were compared for the response to hydrogen peroxide. The different catalytic sources were compared (callus, agar-grown and regularly grown plant tobacco tissues). Better sensitivities with lower variances were obtained from *in vitro*-cultured tissues, which demonstrated the advantages of this approach. Moreover, the storage period of the biosensor was more compared with different plant-tissue-based biosensors.

### 4.2.4 Plant tissue containing acid phosphatase used as a biosensor

PPO and POD are not the main enzymes found in plant tissue such as in potato. Campanella *et al.* [30, 31] used potato tissue as a potential source for acid phosphatase enzyme (AF) and other glucose oxidase enzymes (GODs) to develop a bi-enzymatic biosensor for the detection of phosphate. AF catalyzed the conversion of glucose-6-phosphate to glucose, and it was further oxidized with GOD. Phosphate inhibits the activity of acid phosphatase, and it was proportionate to the concentration of phosphate. The developed biosensor showed a detection limit of up to 62 μM, with a response time of 5 min and storage time of 10–16 days. Furthermore, the developed biosensor was applied in some real samples such as urine, bovine milk, milk powder, red wine, mashed tomatoes and commercial pharmaceutical formulations. The validation of the developed biosensor was also good compared with the standard method.

### 4.2.5 Plant urease-based biosensors

Urease is an enzyme broadly used for the development of urea biosensors. Among the various sources for the enzyme urease, plants were considered the best due to their high titer value and relatively cheap extraction and purification processes.

In this context, soya bean tissue has been used for the development of a biosensor for the detection of urea. Qin *et al.* [32] entrapped soya bean tissue in a miniaturized glass column and developed a flow system to determine the urea in urine samples. The urease hydrolyzes the urea and gives rise to ammonium ion and carbon dioxide. Furthermore, the ion discharged luminol from the column that reacted with the permanganate eluted with NaOH. The reaction between the luminol and permanganate resulted in the development of a luminescence signal. However, the system cannot be strictly referred to as a biosensor due to the noninvolvement of the biocatalyst immobilized on the transducer. The flow injection luminescence analysis is especially attractive, which helps attain high sensitivities and rapid response. In fact, they found a detection limit of 2 µM and an analysis time of 1.5–5 min (including the sampling and washing steps). Moreover, the results were in good agreement with those obtained by standard methods.

## 4.2.6 Plant oxalate oxidase-based biosensors

Oxalate is the metabolic end product in human beings, and the normal range of this end product in urine is 10–30 mg/24 h. Its detection in human urine and plasma is very important from the clinical point of view. Many biosensors have been developed for the detection of oxalic acid based on different bio-components isolated from various sources. The detection of oxalic acid is based on the oxygen or electrons generated from the depletion of hydrogen peroxide. This reaction is catalyzed by the enzyme oxalate oxidase. Spinach (*Spinacia oleracea*) tissue contains a high amount of oxalate oxidase, and it is immobilized on a Clark-type oxygen electrode with albumin. This immobilization was performed by glutaraldehyde cross-linking chemistry onto the electrode surface [33]. The level of oxalate was proportional to the corresponding O consumption during the reaction. The detection limit of the developed biosensor was found to be 10 µM, which is lower than that obtained from bi-enzymatic amperometric (biosensors) or bio-optrode detection systems.

## 4.2.7 Plant tissue containing ascorbic oxidase used as a biosensor

Cucumber (*Cucumis sativus*) tissue homogenate as a potential source of ascorbic acid was used as a biological material for the development of biosensors to detect reduced glutathione (GSH) [34]. The redox behavior of glutathione protects the cells from reactive oxygen species and free radicals, in the transport of amino acids and also in the synthesis of protein and nucleic acid. Cucumber tissue contains a high amount of the ascorbic acid oxidase enzyme, a member of the family of oxidoreductase that catalyzes the oxidization of the water-soluble vitamin ascorbic acid. The developed biosensor showed a detection limit of 0.1 μM. Moreover, its application to real samples like broccoli, potato and tomato provided good results that were comparable to those obtained by the standard method. The storage stability of the biosensor was also extremely good around one month without much loss of significant activity.

### 4.2.8 Plant pyruvate decarboxylase-based biosensors

A biosensor was developed for the detection of aldehydes based on plant tissues containing the enzyme pyruvate decarboxylase and was immobilized onto the optical transducer. Tissue from corn kernel was coupled with a fiber-optic $CO_2$ electrode, where the enzyme catalyzed the conversion of acetaldehyde to carbon dioxide and it was detected by a $CO_2$ electrode. However, most common method to detect $CO_2$ includes the use of the corresponding $CO_2$ membrane electrode. In addition, a fiber-optic $CO_2$optrode may be used if a pH-sensitive absorption indicator is coupled to the device. The results obtained from different electrodes revealed the shorter response time and the somewhat lower detection limits with the optical-based biosensor. The storage stability was about seven days, which is similar to that of the $CO_2$ membrane electrode. It demonstrated that the irradiation from the fiber probes did not damage the tissues or the plant-based materials.

### 4.2.9 Plant tissue containing diamine oxidase used as a biosensor

Both the seed and lentil from leguminous plants contain a high amount of the enzyme diamine oxidase (DAO). DAO catalyzes the oxidation of diamines to produce hydrogen peroxide, which can be measured amperometrically using the principle developed by Botrè *et al.* [35].They developed a biosensor based on pea plant cotyledon to detect the presence of diamines. The detection limit of the developed biosensor was up to 0.5 μM of the diamines. They additionally developed a hybrid biosensor, combining lentil tissue as a source of ornithine decarboxylase enzyme (ODC), to detect ornithine, and lysine decarboxylase enzyme (LDC), to detect lysine contents (Figure 4.3). Analytical performance of the developed hybrid biosensor was almost similar to the measurement using a standard bi-enzymatic electrode. Peroxides were produced by the oxidation of amino acid diamines, and it may be additionally monitored by a luminescence reaction involving luminol and $Co^{2+}$-based chemistry. Mei *et al.* [36] reported packaging of pea-seedling tissue into a mini column, followed by incorporation of the bioreactor into a luminescence flow injection analysis system. The mixture of sensitive luminescence detection with the flow injection analysis system offered numerous advantages compared to a traditional enzyme electrode-based system, like rapid

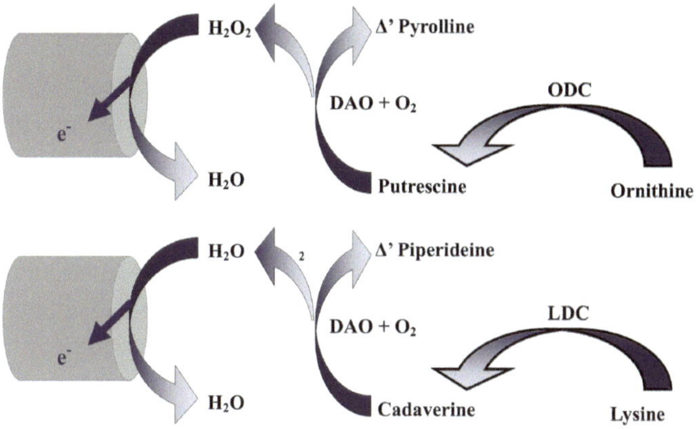


Figure 4.3: Working principle of the legume cotyledon tissue-based biosensors for amino acids.

response, very low detection limit (30 and 60nM for putrescine and cadaverine, respectively), economical and easy to fabricate.

## 4.2.10 Plant tissue containing pectinesterase used as a biosensor

Pectin, a complex carbohydrate with molecular weight ranging between 25,000 and 450,000, is present mainly in the peel of different fruits such as apples, oranges and some vegetables. It is commonly found in fruit jellies and jam. The major disadvantage of earlier reported pectin biosensors was the elevated viscosity of pectin solutions that typically fouls the surface of the working electrode. Biosensors based on plant tissues as the potential source of biocatalyst have shown better results to overcome this drawback because plant tissues can tolerate higher pectin absorption and they are also very cheap and can be easily replaced. Horie and Rechnitz [37] reported the development of a hybrid biosensor for the detection of pectin using mandarin peel tissue as a source of pectin esterase and alcohol oxidase enzymes. The enzyme pectin esterase catalyzed the breakdown of pectin to methanol, which was further oxidized with the enzyme alcohol oxidase to produce hydrogen peroxide. More efforts are needed to detect peroxide. To enhance the system, teflon membrane has been used. In this method, no interference was generated by ascorbic acid, but the response time was quite high due to the slow diffusion of methyl alcohol through the membrane on the working electrode; moreover, the noise was high, and the conductor surface became rough, limiting applications in continuous use. Thus, frequent replacement of the enzyme membrane layer was needed. Despite these problems, they suggested that biosensors based on plant tissues could accept the absorption of substances having a high molecular weight.

### 4.2.11 Plant tissue containing sulfite oxidase for biosensor application

Sulfite is widely used as preservative in the food and beverage industry to inhibit oxidation and microbial growth. The content of sulfite must be strictly monitored due to its harmful effects on hypersensitive people. The plant *Malva vulgaris* has been reported as a potent source of sulfite oxidase (SO) that can be used to develop the biosensor for sulfite detection [38]. SO catalyzes the oxidation of sulfur containing amino acids and produces hydrogen peroxide. The oxygen utilized for the oxidation reaction was monitored by an amperometric transducer, and it was proportional to the sulfite concentration in the samples. Although the tissue-based biosensor conferred a limit of detection (0.2 mM) that was not up to the standard pure enzyme-based biosensors, immobilization protocols were easier and cost effective.

### 4.2.12 Biosensors based on mushroom tissues as the biocatalyst

Mushrooms have unique features and different types, giving them some significant properties and leading them to being placed into an individual category. Mushroom tissues are known to contain a high titer of the PPO enzyme, making them extremely useful as a bio-component for the development of biosensors. Mushroom tissue mixed with banana tissue was immobilized onto an electrochemical transducer system to develop a biosensor for the detection of ortho-diphenol. The analytical performance of the developed biosensor in terms of detection limit was found to be very low and had a rapid response time [39]. Topçu *et al.* [40] together used the PPO content of mushroom tissue for the development of a biosensor to detect phenolic compounds. The tissue was immobilized on a Clark-type oxygen electrode with albumin and glutaraldehyde as the cross-linkers. Except for phenol estimation, an inhibition study was performed. It was also studied as a tool to determine benzoic acid and thiourea in beverages such as soft drinks and fruit juices. Apart from PPO, mushroom tissues also offer many other enzymes, mainly alcohol oxidase. Akylmaz and Dinçkaya [41] investigated the utilization of homogenized tissues of mushroom (*Agaricus bisporus*) as a source of alcohol oxidase that catalyzed the oxidation reaction of alcohol to aldehyde. The presence of aldehyde in alcoholic beverages is harmful, and its detection in the blood serum and in different body fluids is important from the clinical, pharmacology and forensic aspects. The immobilization of tissues homogenizes them onto a transducer with the help of gelatin and glutaraldehyde as the entrapment and the cross-linker, respectively. The developed biosensor provided a rapid response time (2 min) and a wide linear range (from 0.2 to 20 mM).

The high laccase oxidase content in mushrooms can also be used for the biosensor development to detect phenolic compounds. The low stability and poor activity of plant laccase oxidases limited their applications in biosensor development. The employment of plant tissues as a source of enzyme might, in theory, improve the analytical parameters of the biosensor. Timur *et al.* [42] compared the performances of three different biosensors based on enzyme laccase oxidases from totally different sources (laccase from recombinant *Aspergillus niger*, *Trametes versicolor* laccase and *Agaricus bisporus* mushroom tissue). Higher activities were determined at higher pH values with the tissue-based biosensor. This impact, however, may be due to the presence of PPO, which works better at higher pH values, as opposed to laccases.

### 4.2.13 Plant asparaginase-based biosensors

The enzyme asparaginase is widely used as a drug for the prevention and treatment of cancers, especially lympho-sarcoma and acute lymphoblastic leukemia [43]. L- asparaginase enzyme hydrolyzes asparagine into aspartic acid. This enzyme can be produced from a wide range of sources, namely bacteria, algae, fungi, plants, etc. Among them, plants have been found to be better sources because of their good yield, low cost of production and high activity [44]. *Ashwagandha* (*Withania somnifera* (L.) Dunal) is considered a good source of asparaginase and is extensively utilized for the development of a biosensor to detect asparagine [45–47].The developed biosensor was used for the detection of asparagine in food samples such as fruit juices and in clinical samples such as leukemic blood samples. The analytical performance of the developed biosensor was good in terms of storage stability – it was found to be stable for about more than a month if stored at 4 °C; it also had a wide, good linear range of detection from $10^{-1}$ to $10^{-10}$ M.

## 4.3 Conclusions and Future Prospects

The focus of various companies manufacturing diagnostic tools is the development of biosensor devices and their application in various fields. Huge amount of work has been done to develop these biosensors, and while developing them, immobilization of biological materials onto the physicochemical transducers should be ensured. Among various sources, plants and plant-derived materials are extensively used for the development of biosensors because they have major advantages over other biomaterials used in biosensor development. The lab-on-chip technology is also of great interest to investigate plant processes such as growth, development and responses to climatic changes, e.g. change in temperature, air quality and soil salinity. However, further studies are needed in this area to develop practically usable biosensors.

## References


[1] Verma N, Singh A K. Development of BOD biosensor for monitoring the fermentation industry effluents. ISRN Biotechnology 2013, DOI:10.5402/2013/236062.

[2] Singh A K, Verma N. Quartz Crystal Microbalance based approach for food quality. Current Biotechnology 2014, 3, 127–132.

[3] Kumar S, Verma N, Singh A K. Development of Cadmium Specific Recombinant Biosensor and its Application in Milk Samples. Sensors and Actuators B 2017, 240, 248–254.

[4] Verma N, Singh A K, Saini N. Synthesis and characterization of ZnS quantum dots and application for development of arginine biosensor. Sensing and Biosensing Research2017a, 15, 41–45.

[5] Verma N, Singh A K, Singh M. L-arginine biosensor: a comprehensive review. Biochemistry and Biophysical Report 2017b, 12, 128–139.

[6] Campàs M, Carpentier R, Rouillon R. Plant tissue-and photosynthesis-based biosensors. Biotechnology Advances 2008, 26, 370–378.

[7] Narsaiah K, Jha S N, Jaiswal P, Singh A K, Kaur P P, Sharma R, Kumar R. Kinetic Binding Analysis of Biomolecular Interactions Using Surface Plasmon Resonance System. Journal of Agricultural Engineering 2016, 53 (2), 38–44.

[8] Verma N, Kumar S, Kaur H. Fiber Optic Biosensor for the Detection of Cd in Milk. J Biosens Bioelectron 2010, 1,102, doi:10.4172/2155-6210.1000102.

[9] Kaur H, Verma N. High Throughput Optical Biosensor for Monitoring Pb (II) Ions in Milk through Fluorescence based Microarray Approach. J Biosens Bioelectron 2015, 6, 166, doi: 10.4172/2155-6210.1000166.

[10] Ramanathan K, Danielsson B. Principles and applications of thermal biosensors. Biosens Bioeletcron 2001, 16(6), 417–423.

[11] Mishra G K, Mishra R K, Bhand S. Flow injection analysis biosensor for urea analysis in adulterated milk using enzyme thermistor. Biosens Bioeletcron 2010, 26, 1560–1564.

[12] Kuriyama S, Rechnitz GA. Plant tissue-based bioselective membrane electrode for glutamate. Anal Chim Acta 1981,131, 91–6.

[13] Sidwell JS, Rechnitz GA. "Bananatrode" – an electrochemical biosensor for dopamine. Biotech Lett 1985, 7, 419–22.

[14] Wang J, Lin MS. Mixed plant tissue-carbon paste bioelectrode. Anal Chem1988, 60, 1545–8.

[15] Eggins BR, Hickey C, Toft SA, Zhou DM. Determination of flavanols in beers with tissue biosensors. Anal Chim Acta 1997, 347, 281–8.

[16] Cummings EA, Mailley P, Linquette-Mailley S, Eggins BR, McAdams ET, McFadden S. Amperometric carbon paste biosensor based on plant tissue for the determination of total flavanol content in beers. Analyst 1998, 123, 1975–80.

[17] Lima AWO, Nascimento VB, Pedrotti JJ, Angnes L. Coconut-based plant tissue reactor for biosensing of catechol in flow injection analysis. Anal Chim Acta 1997, 354, 325–31.

[18] Fatibello-Filho O, Omuro Lupetti K, Vieira IC. Chronoamperometric determination of paracetamol using an avocado tissue (*Persea americana*) biosensor. Talanta 2001, 55, 685–92.

[19] Felix FS, Yamashita M, Angnes L. Epinephrine quantification in pharmacological formulations utilizing plant tissue biosensors. Biosens Bioelectron 2006, 21, 2283–9.

[20] Mazzei F, Botrè F, Lorenti G, Simonetti G, Porcelli F, Scibona G, et al. Plant tissue electrode for the determination of atrazine. Anal Chim Acta 1995, 316, 79–82.

[21] Wang J, Naser N, Kwon HS, Cho MY. Tissue bioelectrode for organic-phase enzymatic assays. Anal Chim Acta 1992, 264, 7–12.

[22] Vieira IC, Fatibello-Filho O. Biosensor based on paraffin/graphite modified with sweet potato tissue for the determination of hydroquinone in cosmetic cream in organic phase. Talanta 2000, 52, 681–9.
[23] de Oliveira IRWZ, Vieira IC. Immobilization procedures for the development of a biosensor for the determination of hydroquinone using chitosan and gilo (*Solanum gilo*). Enzyme Microbiol Tech 2006, 38, 449–56.
[24] de Oliveira IRWZ, Fernandes SC, Vieira IC. Development of a biosensor based on gilo peroxidase immobilized on chitosan chemically cross linked with epichlorohydrin for determination of rutin. J Pharmaceut Biomed 2006, 41, 366–72.
[25] Fernandes SC, de Oliveira IRWZ, Vieira IC. A green bean homogenate immobilized on chemically cross-linked chitin for determination of caffeic acid in white wine. Enzyme Microb Tech 2007, 40, 661–8.
[26] Oungpipat W, Alexander PW, Southwell-Keely P. A reagentless amperometric biosensor for hydrogen peroxide determination based on asparagus tissue and ferrocene mediation. Anal Chim Acta 1995, 309, 35–45.
[27] Liawruangrath S, Oungpipat W, Watanesk S, Liawruangrath B, Dongduen C, Purachat P. Asparagus-based amperometric sensor for fluoride determination. Anal Chim Acta 2001, 448, 37–46.
[28] Kozan JVB, Silva RP, Serrano SHP, Lima AWO, Angnes L. Biosensing hydrogen peroxide utilizing carbon paste electrodes containing peroxidases naturally immobilized on coconut (Cocus nucifera L.) fibers. Anal Chim Acta 2007, 591, 200–7.
[29] Navaratne A, Rechnitz GA. Improved plant tissue-based biosensor using in vitro cultured tobacco callus tissue. Anal Chim Acta 1992, 257, 59–66.
[30] Campanella L, Cordatore M, Mazzei F, Tomassetti M. Determination of inorganic phosphate in drug formulations and biological fluids using a plant tissue electrode. J Pharmaceut Biomed Anal 1990, 8, 711–6.
[31] Campanella L, Cordatore M, Mazzei F, Tomassetti M, Volpe G. Phosphate determination in foodstuffs using a plant tissue electrode. Food Chem 1992, 44, 291–7.
[32] QinW, Zhang Z, Peng Y. Plant tissue-based chemiluminescence flow biosensor for urea. Anal Chim Acta 2000, 407, 81–6.
[33] Sezgintürk MK, Dinçkaya E. A novel amperometric biosensor based on spinach (*Spinacia oleracea*) tissue homogenate for urinary oxalate determination. Talanta 2003, 59, 545–51.
[34] Sezgintürk MK, Dinçkaya E. An amperometric inhibitor biosensor for the determination of reduced glutathione (GSH) without any derivatization in some plants. Biosens Bioelectron 2004, 19,835–41.
[35] Botrè F, Botrè C, Lorenti G, Mazzei F, Porcelli F, Scibona G. Plant tissue biosensors for the determination of biogenic diamines and of their amino acid precursors: effect of carbonic anhydrase. Sensor Actuat B-Chem 1993,15,135–40.
[36] Mei Y, Ran L, Ying X, Yuan Z, Xin S. A sequential injection analysis/chemiluminescence plant tissue-based biosensor system for the determination of diamine. Biosens Bioelectron 2007,22, 871–6.
[37] Horie H, Rechnitz GA. Hybrid tissue/enzyme biosensor for pectin. Anal Chim Acta 1995,306, 123–7.
[38] Sezgintürk MK, Dinçkaya E. Direct determination of sulfite in food samples by a biosensor based on plant tissue homogenate. Talanta 2005, 65, 998–1002.
[39] Yifeng T. A plant tissue electrode based on the use of chemical mediators. Anal Lett 1993, 26, 1557–66.

[40] Topçu S, Sezgintürk MK, Dinçkaya E. Evaluation of a new biosensor-based mushroom (*Agaricus bisporus*) tissue homogenate: Investigation of certain phenolic compounds and some inhibitor effects. Biosens Bioelectron 2004, 20, 592–7.
[41] Akylmaz E, Dinçkaya E. A mushroom (*Agaricus bisporus*) tissue homogenate based alcohol oxidase electrode for alcohol determination in serum. Talanta 2000, 53, 505–9.
[42] Timur S, Pazarlioglu N, Pilloton R, Telefoncu A. Thick film sensors based on laccases from different sources immobilised in polyaniline matrix. Sensor Actuat B-Chem 2004, 97, 132–6.
[43] Schemer G, Holcenberg JS. Enzyme as Drugs. New York: Wiley Inter Science, 1981, pp. 455–473.
[44] Oza VP, Trivedi SD, Parmar PP, Subramanian RB. *Withania somnifera* (ashwagandha): Novel source of L-asparaginase. J Integ Plant Biol 2009, 51, 201–206.
[45] Kumar K, Kataria M, Verma N. Plant asparaginase-based asparagine biosensor for leukemia. Artificial Cells, Nanomedicine, and Biotechnology 2013, 41, 184–188.
[46] Verma N, Kataria M, Kumar K, Saini J. Comparative study of L-asparaginase from different cytotypes of *Withania somnifera* (L.) Dunal and its purification. J Nat Prod Plant Resour 2012, 2(4), 475–481.
[47] Verma N, Kataria M, Kumar K, Saini J. TEOS hydrosol gel-chitosan matrix based biosensor for monitoring asparagine in various fruit juices. Annals of Biological Research, 2013, 4(2), 276–281.

Pawan Kumar, Pooja Sharma, Vikas Kumar, Amit Kumar, Amit Kumar, Raj Singh and Anil K. Sharma

# 5 Plant resources: *In vitro* production, challenges and prospects of secondary Metabolites from medicinal plants

**Abstract:** The medicinal plant species from the Northwestern Himalayas are highly valued and are utilized in the preparation of pharmaceutical drugs. Phytochemicals/bioactive compounds with therapeutic potential have been procured from high-altitude medicinal plant species for use in the manufacturing of drugs in pharmaceutical industries. *In vitro* production of secondary metabolites attributed with various medicinal properties has been known since three decades. However, the content of metabolites was found to be very low in *in vitro* cultures, limiting their use in meeting the current industrial demands. Information on current *in vitro* metabolite production strategies, biosynthesis and their applications will open up new doors to design novel genetic interventions, including metabolomics, genomics and tissue culture technologies. Knowledge of *in vitro* production, the associated challenges and the prospects of attaining high-value bioactive compounds will provide a platform to not only explore but also gain an understanding of the metabolite biosynthesis to enhance phytochemical yields in various industries.

**Keywords:** Secondary metabolites, medicinal plants, biosynthesis, *in vitro*, pharmacological properties

## 5.1 Introduction

Secondary metabolites are "natural products" mainly produced to aid in the survival of plants under stress conditions. Secondary metabolites can be divided into three groups consisting of 29,000 terpenoids, 12,000 alkaloid derivatives and 8,000 phenolics produced through different biosynthetic routes [1, 2]. Secondary metabolites are used as food additives, drugs, flavors, fragrances, dyes, colors, pesticides, pharmaceuticals, agrochemicals, bio-pesticides and, more recently, as nutraceuticals. It is generally acknowledged that the main problem in enhancing the production of metabolites to higher levels has been the lack of basic knowledge on the biosynthesis, transport and accumulation of metabolites [3]. Traditional approaches including labeled precursor addition, gene

https://doi.org/10.1515/9783110563337-005

cloning, intermediate identification, expressed sequence tag (EST) libraries, enzyme purification and characterization have been used for deciphering the biosynthetic pathway of natural products/metabolites. Such techniques have some limitations like these are very time consuming and difficult to understand the metabolism. The chemistry behind the biosynthesis of metabolites can be revealed by various tools like metabolomics, transcriptomics, proteomics and functional genomics, which can lead to an understanding of secondary metabolite biosynthesis. Secondary metabolites like podophyllotoxin, vinblastine/vincristine, taxol (paclitaxel), artemisinin, shikonin, quinine, picrosides, hypericum, ajmalicine, serpentine and campothecin are being used by the industries for their large-scale production because of their anticancer, antimalarial, hepato-protective, anti-HIV, anti-pyretic, antifungal and anti-inflammatory properties [4, 5]. The world market for herbal drugs has an annual growth rate of between 5 and 15%, with a global market of US $62 billion, which is expected to grow to US $5 trillion by the year 2050. These metabolites have a high demand in international markets, but their supply is limited because of complexity in *in vitro* synthesis, the high cost of production and the decreasing availability of raw materials of medicinal plants. Another reason is that the exact biosynthetic pathway of these metabolites is not known completely or the regulatory steps/genes controlling the metabolite biosynthesis are not known clearly so as to aim at genetic interventions for enhanced metabolite yields. The empirical methods have not been able to meet the increasing world demand for valuable natural products/metabolites, including the anticancer agent paclitaxel, which reached 1040 kg per year in 2013 according to the Global Industry Analysts.

The higher cost of secondary metabolite production and the geographical limitation of medicinal plants have forced researchers to find out alternative methods or sources for the production of these metabolites. Various research groups across the world have been working toward the enhancement of secondary metabolite production in *in vitro* cultures, with very limited success. Therefore, there exists an urgent need to not only decipher biosynthetic pathways but also identify key points in the biosynthesis, transport and accumulation of metabolites [6, 7]. Despite tremendous efforts, a genetically stable transformed medicinal plant has not been achieved to the best of our knowledge. Metabolic engineering requires a basic understanding of metabolic networks and their regulation, identification of the accurate regulatory genes/enzymes, subcellular localization of the enzymes, metabolic channels and transporters involved in the biosynthetic pathways [8], and due to lack of appropriate knowledge in the above-mentioned factors, metabolic engineering often results in low yields of metabolites and undesirable products.

Recent studies have shown that various transporters and channels exist in plants for the transport and accumulation of metabolites. However, little understanding exists regarding the genes involved in transportation and accumulation vis-à-vis metabolite production in various tissues, which could lead to an increase in the productivity of valuable secondary metabolites. Modern genomic technologies such as whole-genome transcriptomics, proteomics and metabolomics coupled with metabolic network analysis are expected to uncover several gaps in the biology of metabolite biosynthesis and accumulation [4, 9, 10]. Currently, high-throughput *de novo* sequencing of the transcriptome of various valuable medicinal plants, such as *Podophyllum hexandrum, P. peltatum, Picrorhiza kurroa, Artemisia annua, Hypericum perforatum, Catharanthus roseus, Atropa belladonna, Taxus brevifolia, Swertia chirata, Valeriana officinalis, Rauvolfia serpentine, Digitalis lanata, Linum flavum, Datura innoxia, Acacia aulacocarpa, Aglaia foveolata Pannell, Andrographis paniculata, Calophyllum antillanum, Camptotheca acuminata, Castanospermum australe A., Cephalotaxus harringtonii, Cinchona pubescens, Colchicum autumnale, Coleus forskohlii, Datura ceratocaula, Datura metel, Datura stramonium L., Datura wrightii Regel, Galanthus elwesii, Gloriosa superba, Huperzia lucidula, Huperzia squarrosa, Hyoscyamus niger, Larrea tridentata, Linum usitatissimum, Narissus pseudonarcissus, Papaver somniferum, Pilocarpus pennatifolius, Pimpinella anisum, Plectranthus caninus, Strophanthus gratus, Taxus x media* and *Tripterygium regelii*, provides a platform for rapid gene identification involved in the biosynthesis of a particular metabolite. Combining the transcriptomics, proteomics and metabolomics approaches together can allow a better understanding of metabolite biosynthesis and its regulation [4, 9, 10].

We have, therefore, reviewed the *in-vitro* production, challenges and prospects of high value secondary metabolites such as podophyllotoxin, with anosides and shikonin procured from high altitude medicinal plant species. Knowledge of these important metabolites will provide a useful resource to industries and the scientific community to better explore and gain an understanding of secondary metabolite biosynthesis.

## 5.2 *In-vitro* production of secondary metabolites

Different kinds of secondary metabolites, namely alkaloids, terpenoids and alkaloids, have been produced *in vitro* utilizing cell culture techniques/interventions. Bioactive compounds such as shikonin, podophyllotoxin, withanosides, vincrestine, taxol, catharanthine, vindoline, ajmalicine and anthocyanin are some of the important examples of secondary metabolites that have been commercialized [11, 12] (Figure 5.1).

Various *in vitro* production strategies have been employed in the past, such as precursor feeding, modification in production media, elicitor's treatments, immobilization of cells and scaling up through bioreactors [3, 4, 6, 13, 14]. A few important secondary metabolites produced in *in vitro* conditions are listed in Table 5.1.

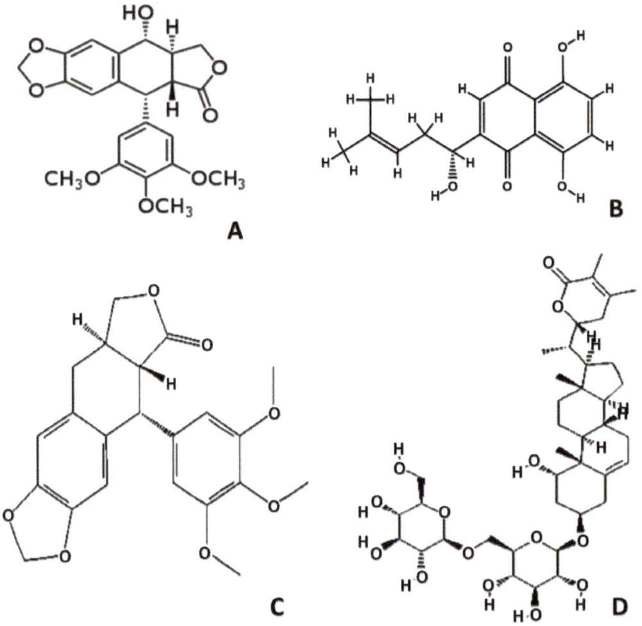

Figure 5.1: Structures of bioactive compounds: (A) podophyllotoxin, (B) shikonin, (C) deoxypodophyllotoxin and (D) withanosides.

Table 5.1: Secondary metabolite contents observed in different medicinal plants and fungusspecies (*in vitro* conditions).

| Plant Family | Medicinal plants | Podophyllotoxin content (% of dry weight) | Plant parts | References |
|---|---|---|---|---|
| ***Berbediaceae*** | *Podophyllum hexandrum* | 4.3 | Roots/rhizomes | [3,6] |
| | *P. peltatum* | 0.25 | Shoots, roots/rhizomes | [6] |
| | *P. versipelle* | 0.32 | Roots/rhizomes | [15] |
| | *Diphylleia grayi* | 1.3 | Roots | [15] |
| | *Diphylleia cymosa* | 0.54 | Leaves | [15] |
| ***Linaceae*** | *Linum album* Kotschy ex Boiss | 0.35 | Callus/suspension cultures | [16] |
| | *Linum nodiflorum* | 1.7 | Callus/suspension cultures | [17] |
| | *L. flavum* | 0.2 | Callus/suspension cultures | [18] |
| | *L. flavum* | 0.16 | Roots | [19] |

Table 5.1 (continued)

| Plant Family | Medicinal plants | Podophyllotoxin content (% of dry weight) | Plant parts | References |
|---|---|---|---|---|
| ***Cupressaceae*** | *Juniperus chinensis* | 0.0025 | Leaves (needles) | [20] |
| | *J. chinensis* | 0.005 | Callus/suspension cultures | [20] |
| | *J. scopulorum* | 0.17 | Leaves (needles) | [21] |
| | *J. sabina* | 0.20 | Leaves (needles) | [21] |
| | *J. lucayana* | 0.10 | Leaves (needles) | [21] |
| | *J. virginiana* | 0.10 | Leaves (needles) | [21, 22] |
| | *J. silicicola* | 0.04 | Leaves (needles) | [21] |
| | *J. thurifera* | 1.3 | Leaves (needles) | [23] |
| | *Callitris drummondii* | 1.4 | Leaves (needles) | [24] |
| ***Polygalaceae*** | *Polygala polygama* | 0.08 | Whole plant | [25] |
| **Fungus species** | *Trametes hirsute* | 0.03 | Biomass | [26] |
| | *Fusarium oxysporum* | 0.02 | Biomass | [26] |
| ***Boraginaceae*** | *Lithospermum erythrorhizon* | 1–2.5 µg/mL (shikonin content) | Dried roots, root hairs | [27, 28, 29] |

## 5.2.1 Elicitation approaches for the enhancement of secondary metabolite contents

Secondary metabolites were isolated from their natural plant resources including roots, rhizome, shoots, bark and flower because their chemical synthesis is quite complex and costly. The contents of *in vitro* compounds generated from different medicinal plant species are very low. Metabolite contents can be increased through elicitation approaches by adding elicitor's *in vitro* techniques [30]. The metabolites content in the cell suspension culture of *Ammi majus* L., *Panax ginseng, Rubia akane* were elicited with benzo (1,2,3)-thiadiazole-7-carbothionic acid S-methyl ester, selenium and chitosan (0.5 mM) [31–33]. An *in vitro* production of solasodine with sodium chloride (150 mM) in the cell culture of *Solanum nigrum* was carried out [34]. Similarly, hyoscyamine/scopolamine production was elicited 3–20 times by the addition of $KNO_3$ (35 mM) in a tissue culture of *Atropa belladonna* L. [35]. Silymarin production was increased by methyl jasmonate (10 µM) in *Silybum marianum* tissue culture [36].

### 5.2.2 Metabolic engineering in medicinal plant species

Metabolic engineering is an alternative way for optimizing genetic and regulatory processes to attain the desired amount of natural products from medicinal plants. Many plant species such as *Nicotiana tabacum* (tobacco), *Atropa belladonna, Artemisia annua, Catharanthus roseus* (Madagascar periwinkle) and *Digitalis lanata* have been genetically engineered to enhance the metabolite content [37]. In the past few years, different genetic transformation technologies have been applied for DNA delivery into the host cells like insertion of genes either indirectly, via genetic vectors such as *Agrobacterium tumefaciens* and *A. rhizogenes*, or directly, such as through a particle gun, protoplast fusion, electroporation and microinjection, with successful increase in metabolite production [38]. Recently, new trends are being used in metabolic engineering like heterologous expression, metabolic flux analysis, RNA interference technologies (RNAi) and overexpression analysis of genes involved in the biosynthetic pathways, which aim to achieve a highly efficient productive complex. One interesting example of genetic engineering has been found in *Catharanthus roseus* in which strictosidine synthase (Str) has been overexpressed to achieve higher metabolite production [39]. Similarly, hyscyamine 6β-hydroxylase has been overexpressed in *Hyoscymus muticus* to increase the production of scopolamine substantially [40].

### 5.2.3 Pharmacological applications

In this chapter, we have summarized the important pharmacological properties of some bioactive compounds, namely withanosides, shikonin and podophyllotoxin.

#### 5.2.3.1 Withanosides

*Withania somnifera* is an important medicinal herb having numerous medicinal properties attributed to the presence of secondary metabolites or bioactive compounds. This herb, which is commonly known as *ashwagandha*, or Indian ginseng, is distributed in different regions of the Northwestern Himalayas. Different parts (roots and rhizomes) of Withania species have been utilized in the preparation of many drugs in several industries. This herb exhibits various medicinal properties such as anti-inflammatory activity, anti-stress activity, antibiotic activity, antiparkinsonian properties, antiaging activity and antioxidant effects. The *in vitro* production of bioactive compounds from this herb is very low in tissue culture conditions. The global increasing demands for secondary metabolites from Withania species have led the scientific community to find ways to further increase the *in vitro* metabolite contents.

**Anti-inflammatory activity**

Bioactive compounds of *Withania somnifera* possess anti-inflammatory activity via inhibition of 5-hydroxytryptamine, prostaglandins and histamine. Anti-inflammatory activity was particularly shown by Withaferin A, important bioactive compounds procured from Withania species. Withaferin A is a primary compound that can inhibit arthritic syndrome with least toxic effects. Many studies have shown the effect of *Withania somnifera* in *in-vivo* models, with significant results of different anti-inflammatory activities such as cotton pellet granuloma, adjuvant-induced arthritis and carrageenan-induced inflammation [41].

**Anti-stress properties**

Since a decade, *Withania somnifera* has been known for its anti-stress properties, attributed to the presence of bioactive compounds. This activity of *Withania somnifera* was demonstrated by Kaur *et al.* [42] in the C-H-R model (cold, hypoxia, restrain) by applying bioactive constituents (20 mg/Kg bw) isolated from the roots of *Withania somnifera* in rats. Highly significant results were obtained from the designed experimental model as the rats were more resistant to stress conditions. Numerous studies across the globe have highlighted the importance of *Withania somnifera* in relation to its anti-stress properties.

**Antibiotic activity**

Besides being of interest to the scientific community because of its pharmaceutical properties, *Withania somnifera* – a high-value medical plant species – also has an important medicinal property, namely its antibiotic activity. Its antibiotic activity was highlighted by Bisht and Rawat in 2014, where they used Withania leaf extract against the *Staphylococcus aureus* and *Enterococcus* species, and it was found that 2 mg/mL (100 µL) concentration inhibited the growth of microorganisms with an average zone of inhibition of 20.6 mm and 19.4 mm, respectively [43]. Different parts of *Withania somnifera* have been utilized for examining its antibiotic activity. The active constituents of Withaferin A (10 µg/mL) from Withania species suppressed the enzymatic activity of glucose-6-phosphatedehydrogenase in *Bacillus subtilis*. The antibiotic property of Withaferin A is due the presence of a lactone ring in its structure. The antibiotic activity of Withania species was also validated in a rabbit model that showed a higher activity compared to penicillin [44].

**Antioxidant effect**

The antioxidant capacity of the therapeutic metabolite, fistein, has been explored recently for its role in scavenging free radicals, which can alter molecular components such as amino acids, nucleic acids, lipids and carbohydrates. A proper balance between free radicals and antioxidants is needed for proper physiological functioning;

an imbalance might cause oxidative stress, which leads to numerous diseases such as mutagenesis, diabetes mellitus, neurodegenerative diseases, retinal degeneration, aging and carcinogenesis. Bioactive compounds from *Withania somnifera* were also tested for antioxidant potential via determining the free-radical scavenging enzymes like glutathione peroxidase, catalase and superoxide dismutase [44]. Different tissues/ parts of Withania species have been utilized for studying their antioxidant activities through 1,1-diphenyl-2-picrylhydrazyl (DPPH)2,2′-azino-bis(3-ethylbenzothiazoline-6-sulphonic acid) (ABTS) and Ferric Reducing Antioxidant Power (FRAP) methods.

**Antiaging activity**

*Withania somnifera* extracts have antiaging-inducing properties. Aging is related to the telomerase activity in organisms, and hence maintenance of telomerase is essential to delay the aging process. Bioactive compounds from *Withania somnifera* are known for their antiaging potential by increasing the telomerase activity. The telomerase rapid amplification protocol (TRAP) assay is used to determine the antiaging potential of *Withania somnifera* in the human HeLa cell lines. A significant increase (45%) was reported in the telomerase activity in HeLa cell lines at 10–50 µg conc. of Withania root extracts [45].

**Antiparkinsonian properties**

Parkinson's disease is a type of neurodegenerative disorder identified by the selective loss of dopamine (DA) neurons. One of the most commonly used animal models for Parkinson's disease is 6-hydroxydopamine (6-OHDA). A huge wealth of evidence is available in the literature supporting the role of 6-OHDA in toxic manifestations through oxidant stress. Antiparkinsonian properties of Withania extract have been widely studied due to the presence of bioactive compounds/ metabolites. In a research conducted on rats for five weeks, Withania extract was orally given for 3 weeks to investigate its antiparkinsonian potential. The neurobehavior of rats was checked by determining the levels of glutathione reductase, glutathione-S-transferase, catecholamine content, superoxide dismutase, etc. on the 21st day. This study highlighted the importance of *Withania somnifera* for its antiparkinsonian properties [46].

**Anticarcinogenic activity**

Withania species have been known to have anticarcinogenic properties attributed to the high-value biomolecules withanolides, withaferin and their derivatives. Different types of tissues/parts such as roots, leaves and bark have been used to examine the anticancer activity of Withania species [47]. Bioactive compounds of the medicinal herb *Withania somnifera* are responsible for its anticancer potential as has been revealed in many published reports. Withania extract suppresses the nuclear factor kappa B as well as the necrosis factor in cancer cell lines and hence has been suggested for use in controlling the proliferation of cancer [47]. Different types of cancers, including lung cancer, blood cancer and brain cancer, have been treated with the bioactive constituents of Withania species. Many herbal formulations having anticancer properties prepared using Withania species are available in the market.

#### 5.2.3.2 Shikonin

Shikonin, a red napthoquinone pigment, is extracted from *Lithospermum erythrorhizon*. It is biosynthetically derived from two key precursors, 4-hydroxy benzoate (4HB) and geranyldiphosphate (GPP). It is a secondary metabolite known for its different biological activities such as antimicrobial, insecticidal, antitumor and antioxidants. These compounds are usually colored pigments and therefore have varied applications in food, textiles and cosmetics. Shikonin and its derivatives are distributed among members of the family Boraginaceae, which are commercially valuable for the production of naphthoquinone pigments isolated from roots; they are known to exert wound-healing effects. It includes different species of *Lithospermum, Arnebia, Alkanna, Anchusa, Echium* and *Onosma* Figure 5.2. There are various methods to increase the levels of shikonins in plant cells, such as selection of cell lines, optimization of culture conditions, elicitation, *in situ* product removal, genetic transformation and metabolic engineering.

**Anti-inflammatory activity**

In autoimmune hepatitis disease, several inflammatory cytokines (IL-1$\beta$, TNF-$\alpha$, IFN-$\gamma$ and IL-6) are released, which are the main causes of liver injury [48, 49]. Especially cytokine IL-1$\beta$ that is released by macrophages plays a significant role in the necrosis of liver tissues [50]. Shikonin acts as an anti-inflammatory molecule and downregulates the expressions of the inflammatory cytokines TNF-$\alpha$, IFN-$\gamma$ and IL-1$\beta$. The levels of serum aminotransferase (AST), alanine aminotransferase (ALT), hepatocyte apoptosis and necrosis of liver tissues significantly reduced with the accurate shikonin dose [51]. Shikonin also suppressed the c-Jun N-terminal kinase (**JNK**) signaling pathway, which is activated by immune responses, cell stress and inflammatory cytokine. Shikonin could also prevent liver tissues from injury by inhibiting the expression of interleukin and the phosphorylation of JNK. Shikonin can also be helpful in relieving symptoms of rheumatoid arthritis by suppressing the proliferation of synovial fibroblasts and inhibiting cytokine expression and production [52].

### Antitumor activity

Shikonin and its derivatives were initially observed in *in vitro* cytotoxicity against cancerous cells in 1974 in the United States The anticancer activity of shikonin was closely associated with the promotion of apoptotic death. Several *in vitro* and *in vivo* studies have demonstrated the antitumor activities of shikonin and its derivatives toward various types of cancer cells, such as leukemia cells, breast cancer cells, glioma cells, bladder cancer cells and lung cancer cells. Shikonin functions as a proteasome inhibitor and topoisomerase-I inhibitor, which helps in cancer treatment. In addition, shikonin exerts neuroprotection activity by inhibiting cell apoptosis and oxidative stress [53].

### Antioxidant activity

Shikonin can decrease toxicity from living organisms owing to its radical scavenging properties. It can provide protection against neurodegradation and hepatic ischemia by reducing the reactive oxygen species (ROS) [54]. Similarly, toxicity in PC12 cells due to ROS can be gradually reduced with the help of shikonin. However, shikonin could improve the antioxidant activity of ROS by increasing the concentration of specific enzymes such as superoxidedismutase, catalase and glutathione peroxidase [54].

### Antimicrobial activity

Shikonin exhibits *in vitro* antibacterial activities against both methicillin-sensitive and methicillin-resistant *S. aureus*. Subsequently, shikonin ointment was proved to exhibit an antibacterial effect against *S. aureus* in open wounds in rats [55]. Shikonin has stable fungi-static effects against various cultures of the genus candida and *Trichosporon* genus. Propionyl shikonin and b-hydroxy isovaleryl shikonin, isolated from the roots of *Lithospermum erythrorhizon*, showed both antifungal (*Cladosporium herbarum*) and antiviral (tobacco mosaic virus) activities. It was observed that shikonin and deoxyshikonin have much stronger activities than fluconazole against yeast-like fungi (*Candida kcrusei, Saccharomyces cerevisiae* and *C. glabrata*).

Shikonin, which is an effective natural antibiotic, has the potential to substantially reduce the massive use of existing antibiotic doses. However, shikonin may be important for understanding the stochastic mechanism behind the antibacterial activity in natural compounds.

### Wound-healing activity

Shikonin has natural wound-healing activities at the molecular and cellular levels by controlling the cell trans-differentiation processes and microRNA regulation [56]. Shikonin extracted from the root bark of *Onosma echioides* is found to promote the wound-healing process [57].

#### 5.2.3.3 *Podophyllum hexandrum* as a potential source of podophyllotoxin

*Podophyllum hexandrum* Royle (syn. *P. emodi* Wall), which is native to the alpine and subalpine regions of Himalayan countries like India, Nepal, Pakistan, China, Afghanistan and Bhutan (2500–4200 m amsl), is commonly known as "bankakri or Giriparpat" in the Western Himalayan region (Figure 5.2). *P. hexandrum* Royle has a high level of podophyllotoxin content (4.3%) compared to other podophyllotoxin-producing medicinal plants such as North American *Podophyllum peltatum* (0.25%), *Linum album* (0.35%), *Juniperus chinensis* (0.0025%), *Linum nodiflorum* (1.7%) and *Callitris drummondii* (1.4%). Currently, the roots and rhizome of *P. hexandrum* Royle are the main sources of podophyllotoxin. The rhizome of *P. hexandrum* is preferred more over *P. peltatum* because its resin is more copious and richer in podophyllotoxin, with a content of about 4.3% dry weight as against 0.25% in *P. peltatum* [4]. Although it is possible to chemically synthesize podophyllotoxin, this process has not yet been optimized to be economically feasible. Consequentially, the rhizomes of *P. hexandrum* have been indiscriminately collected to meet the ever-increasing demands for the drug and its derivatives. This over-collection has led to severe habitat destruction, followed by an acute depletion in the population of this herb. The cost of production, lack of organized cultivation, long juvenile phase, poor fruit-setting ability, seed dormancy, overexploitation and geographical limitation of *Podophyllum species*, all forced researchers to find other methods or sources to address the ever-increasing demands of podophyllotoxin in the industries.

Figure 5.2: Medicinal plant species in their natural habitats: (a) *Podophyllum hexandrum* Royle (Indian/Himalayan May Apple), (b) *Podophyllum peltatum* (North American May Apple) and (c) *Arnebia euchroma* (Ratanjot).

## 5.3 Conclusions

Bioactive compounds of immense value have been procured from diverse medicinal plant species across the globe. The demands for high-value metabolites/phytochemicals are gradually escalating, but their *in vitro* production is very low in tissue culture conditions. Genetic engineering is the only way to enhance the *in vitro* metabolite content and fulfill the industrial demands. Knowledge of the *in vitro* production, challenges and prospects of these bioactive compounds could prove useful in developing a new way to explore high-value medicinal plant species. Comprehensive information of metabolites/phytochemicals will be very helpful to the scientific community for further research in this field.

## References


[1] Kumar P, Kumar V, Garlapati VK. Biosynthesis and pharmacological evaluation of shikonin-A highly valuable metabolite of North-Western Himalayas: mini review. Medicinal Pl-Inter J of Phytomed and Related Industries 2016, 8(4), 267–74.

[2] Zwenger S, Basu C. Plant terpenoids: applications and future potentials. Biotechn and Molecular Biol. Reviews 2018, 3, 1–7.

[3] Kumar P, Pal T, Sharma N, Kumar V, Sood H, Chauhan RS. Expression analysis of biosynthetic pathway genes vis-à-vis podophyllotoxin content in *Podophyllum hexandrum* Royle. Protoplasma 2015, 252(5), 1253–1262.

[4] Kumar P, Padhan JK, Kumar A, Chauhan RS. Transcriptomes of *Podophyllum hexandrum* unravel candidate miRNAs and their association with the biosynthesis of secondary metabolites. J Pl Biochem and Biotech 2018, 27(1), 46–54.

[5] Kumar P, Saini M, Bhushan S, Warghat AR, Pal T, Malhotra N, Sood A. Effect of salicylic acid on the activity of PAL and PHB geranyltransferase and shikonin derivatives production in cell suspension cultures of *Arnebia euchroma* (Royle) Johnst – A medicinally important plant species. Appl biochem and biotechn 2014, 173(1), 248–258.

[6] Kumar P, Jaiswal V, Pal T, Singh J, Chauhan RS. Comparative whole-transcriptome analysis in Podophyllum species identifies key transcription factors contributing to biosynthesis of podophyllotoxin in *P. hexandrum*. Protoplasma. 2017, 254(1), 217–228.

[7] Kumar P, Sharma R, Jaiswal V, Chauhan RS. Identification, validation, and expression of ABC transporters in *Podophyllum hexandrum* and their role in podophyllotoxin biosynthesis. Biologia plantarum 2016, 60(3), 452–458.

[8] Peters RJ, Croteau RB. Metabolic engineering of plant secondary metabolism. Handbook of plant biotechnology 2004.

[9] Pal T, Padhan JK, Kumar P, Sood H, Chauhan RS. Comparative transcriptomics uncovers differences in photoautotrophic versus photoheterotrophic modes of nutrition in relation to secondary metabolites biosynthesis in *Swertia chirayita*. Molecular biology reports 2018, 18, 1–22.

[10] Padhan JK, Kumar P, Sood H, Chauhan RS. Prospecting NGS-transcriptomes to assess regulation of miRNA-mediated secondary metabolites biosynthesis in *Swertia chirayita*, a medicinal herb of the North-Western Himalayas. Medicinal Plants-International J Phytomedicines and Related Industries 2016, 8(3), 219–228.

[11] Onrubia M, Moyano E, Bonfill M, Cusidó RM, Goossens A, Palazón J. Coronatine, a more powerful elicitor for inducing taxane biosynthesis in *Taxus media* cell cultures than methyl jasmonate. J Plant Physiol 2013, 170, 211–19.

[12] Smith SL, Slywka GW, Krueger RJ. Anthocyanins of *Strobilanthes dyeriana* and their production in callus culture. J Nat Prod 1981, 44, 609–10.

[13] Chattopadhyay S, Mehra RS, Srivastava AK, Bhojwani SS, Bisaria VS. Effect of major nutrients on podophyllotoxin production in *Podophyllum hexandrum* suspension cultures. Appl Microbiol and Biotech 2003, 60, 541–546.
[14] Anrini M, Jha S. Characterization of podophyllotoxin yielding cell lines of *Podophyllum hexandrum*. Caryologia, 2009, 62 (3), 220–235.
[15] Broomhead AJ, Dewick PM. Tumor inhibitory aryltralin lignans in *Podophyllum versipelle, Diphyllcia cymosa* and *Diphyllicia grayi*. Phytochemistry, 1990, 2, 3831–3837.
[16] Yousefzadi M, Sharifi M, M. Behmanesh, E. Moyano, and J. Palazon, "Salicylic acid improves podophyllotoxin production in cell cultures of *Linum album* by increasing the expression of genes related with its biosynthesis", Biotechnol Lett. 2010, 32, 1739–1743.
[17] Kuhlmann S, Kranz K, Lücking B, Alfermann A, Petersen M. Aspects of cytotoxic lignan biosynthesis in suspension cultures of *Linum nodiflorum*. Phytochemistry Reviews, 2002, 1(1) 37–43.
[18] Smollny T, Wichers H, Kalenberg S, Shasavari A, Petersen M, Alfermann AW. Accumulation of podophyllotoxin and related lignans in cell suspension cultures of *Linum album*. Phytochemistry, 1998, 48, 975–979.
[19] Broomhead AJ, Dewick PM. Aryltetralin lignans from *Linum flavum* and *Linum capitatum*. Phytochemistry 1990, 29, 3839–3844.
[20] Muranaka T, Miyata M, Ito K, Tachibana S. Production of podophyllotoxin in *Juniperus Chinensis* Callus cultures treated with oligosaccharides and a biogenetic precursor in honour of professor G. H. Neil Towers 75th birthday, Phytochemist, 1998, 49(2), 491–496.
[21] Hertwell JL, Johnson JM, Fitzgerald DB, Belkin M. Podophyllotoxin from *Jiniperus* species: Savinin. J Am Chem Soc 1953, 75, 235–236.
[22] Vogt T. Phenylpropanoid biosynthesis. Molecular Pl 2010, 3, 2–20.
[23] San Feliciano A, Miguel del Corral JM, Gordaliza M, Castro MA. Lignans from *Juniperus thruifera*. Phytochemistry 1989, 28, 2863–2866.
[24] Van Uden W, Pras N, Vossebeld E, Mol J, Malingré T. Production of 5-methoxypodophyllotoxin in cell suspension cultures of *Linum flavum* L., Pl Cell Tiss Organ Cult 1990, 20(2) 81–87.
[25] Hokanson GC, Podophylltoxion and 4-demethylpodophyllotoxin from *Polygala polygama* (Polygalaceae), Lioyodia 1979, 41, 497–498.
[26] Yousefzadi M, Sharifi M, Behmanesh M, Moyano E, Bonfill M, Cusido R, Palazon J. Podophyllotoxin: Current approaches to its biotechnological production and future challenges. Eng Life Sci 2010, 10(4), 281–292.
[27] Chen X, Yang L, Zhang N, Turpin JA, Buckheit RW, Osterling C, Oppenheim JJ, Howard OM. Shikonin, a component of chinese herbal medicine, inhibits chemokine receptor function and suppresses human immunodeficiency virus type 1. Antimicrob Agents Chemother 2003, 47 (9), 2810–6.
[28] Tatsumi K, Yano M, Kaminade K, Sugiyama A, Sato M, Toyooka K, Aoyama T, Sato F, and Yazaki K. Characterization of Shikonin Derivative Secretion in *Lithospermum erythrorhizon* Hairy Roots as a Model of Lipid-Soluble Metabolite Secretion from Plants. Front Plant Sci 2016, 7, 1066.
[29] Yeh YC, Liu TJ, Lai HC. Shikonin Induces Apoptosis, Necrosis, and Premature Senescence of Human A549 Lung Cancer Cells through Upregulation of p53 Expression. Evidence-Based Complementary and Alternative Medicine 2015. http://dx.doi.org/10.1155/2015/620383
[30] Namdeo AG. Plant cell elicitation for production of secondary metabolites: A review. Pharmacognosy Rev 2007, 1, 69–79.
[31] Staniszewska I, Krolicka A, Malinski E, Lojkowska E, Szafranek J. Elicitation of secondary metabolites in in vitro cultures of *Ammi majus* L. Enzyme Microb Technol 2003, 33, 565–568.

[32] Jeong GA, Park DH. Enhanced secondary metabolite biosynthesis by elicitation in transformed plant root system. Appl Biochem Biotechnol 2007, 130, 436–446.
[33] Jin JH, Shin JH, Kim JH, Chung IS, Lee HJ. Effect of chitosan elicitation and media components on the production of anthraquinone colorants in madder (Rubia akane Nakai) cell culture. Biotechnol Bioprocess Eng 1999, 4, 300–304.
[34] Bhat MA, Ahmad S, Aslam J, Mujib A. Salinity stress enhances production of solasodine in Solanum nigrum L. Chemical and Pharmaceutical Bulletin 2008, 56(1), 17–21.
[35] Chashmi NA, Sharif M, Karimi F, Rahnama H. Differential production of tropane alkaloids in hairy roots and in vitro cultured two accessions of *Atropa belladonna* L. 6under nitrate treatments. Z Naturforsch C 2010, 65, 373–9.
[36] AbouZid S. Yield improvement strategies for the production of secondary metabolites in plant tissue culture: silymarin from *Silybum marianum* tissue culture. Nat Prod Res 2014, 20, 1–9. doi: 10.1080/14786419.2014.927465
[37] Atanasov AG, Waltenberger B, Pferschy-Wenzig EM, Linder T, Wawrosch C, Uhrin P, Temml V, Wang L, Schwaiger S, Heiss EH, Rollinger JM. Discovery and resupply of pharmacologically active plant-derived natural products: a review. Biotechn advan 2015, 33(8), 1582–614.
[38] Gelvin SB. *Agrobacterium*-mediated plant transformation: the biology behind the "gene-jockeying" tool. Microbiology and molecular biology reviews 2003, 67(1), 16–37.
[39] Canel C, Lopes-Cardoso MI, Whitmer S, van der Fits L, Pasquali G, van der Heijden R, Hoge JH, Verpoorte R. Effects of over-expression of strictosidine synthase and tryptophan decarboxylase on alkaloid production by cell cultures of *Catharanthus roseus*. Planta 1998, 205(3), 414–9.
[40] Zhang L, Ding R, Chai Y, Bonfill M, Moyano E, Oksman-Caldentey KM, Xu T, Pi Y, Wang Z, Zhang H, Kai G. Engineering tropane biosynthetic pathway in Hyoscyamus niger hairy root cultures. Proceedings of the national academy of sciences of the United States of America 2004, 27, 101(17), 6786–91.
[41] Jayaprakasam B, Zhang Y, Seeram N Nair. Growth inhibition of human tumor cell lines by withanolides from *Withania somnifera* leaves. M Life Sci 2003, 74, 125–32.
[42] Kaur P, Mathur S, Sharma M, Tiwari M, Srivastava KK, Chandra R. A biologically active constituent of *withania somnifera* (ashwagandha) with antistress activity. Ind J Clinical Biochem. 2001, 16(2), 195–8.
[43] Bisht P, Rawat V. Antibacterial activity of *Withania somnifera* against Gram-positive isolates from pus samples. Ayu. 2014, 35(3), 330.
[44] Sumathi S, Padma PR, Gathampari S, Vidhya S. Free radical scavenging activity of different parts of *Withania somnifera*. Anc Sci of Life 2007, 26(3), 30.
[45] Raguraman V, Subramaniam JR. *Withania somnifera* root extract enhances telomerase activity in the human hela cell line. Advances in Bioscience and Biotech. 2016, 15, 7(04), 199.
[46] Ahmad M, Saleem S, Ahmad AS, Ansari MA, Yousuf S, Hoda MN, Islam F. Neuroprotective effects of *Withania somnifera* on 6-hydroxydopamine Induced Parkinsonism in rats. Hum Exp Toxicol. 2005, 24(3), 137–147.
[47] Rai M, Jogee PS, Agarkar G, Santos CA. Anticancer activities of *Withania somnifera*: Current research, formulations, and future perspectives. Pharmaceutical biol. 2016, 54(2), 189–97.
[48] Tiegs G. Cellular and cytokine-mediated mechanisms of inflammation and its modulation in immune-mediated liver injury. Z Gastroenterol 2007, 45(1), 63–70
[49] Wang HX, Liu M, Weng SY, Li JJ, Xie C, He HL, Guan W, Yuan YS, Gao J. Immune mechanisms of Concanavalin A model of autoimmune hepatitis. World J Gastroenterology 2012, 18(2), 119–125
[50] Cui K, Yan G, Xu C, Chen Y, Wang J, Zhou R, Bai L, Lian Z, Wei H, Sun R. Invariant NKT cells promote alcohol-induced steatohepatitis through interleukin-1$\beta$ in mice. J Hepatol 2015, 62(6), 1311–1318.

[51] Liu T, Xia Y, Li J, Li S, Feng J, Wu L, Zhang R, Xu S, Cheng K, Zhou Y, Zhou S, Dai W, Chen K, Wang F, Lu J, Zhou Y, Guo C. Shikonin Attenuates Concanavalin A-Induced Acute Liver Injury in Mice *via* Inhibition of the JNK Pathway. Mediators of Inflammation 2016. http://dx.doi.org/10.1155/2016/2748367

[52] Yang KY, Chen DL. Shikonin Inhibits Inflammatory Response in Rheumatoid Arthritis Synovial Fibroblasts *via* lncRNA-NR024118. Evid Based Complement Alternat Med 2015.

[53] Tong Y, Bai L, Gong R, Chuan J, Duan X, Zhu Y. Shikonin Protects PC12 Cells Against β-amyloid Peptide-Induced Cell Injury through Antioxidant and Antiapoptotic Activities. Scientific Reports 2017, 8. Doi:10.1038/s41598-017-18058-7

[54] Wada N, Kawano Y, Fujiwara S, Kikukawa Y, Okuno Y, Tasaki M, Ueda M, Ando Y, Yoshinaga K, Ri M, Iida S, Nakashima T, Shiotsu Y, Mitsuya H, Hata H. Shikonin, dually functions as a proteasome inhibitor and a necroptosis inducer in multiple myeloma cells. In J oncology 2015, 46, 963–972.

[55] Lee YS, Lee DY, Kim YB, Lee SW, Cha SW, Park HW, Kim GS, Kwon DY, Lee MH, Han SH. The Mechanism Underlying the Antibacterial Activity of Shikonin against Methicillin-Resistant Staphylococcus aureus. Evid Based Complement Alternat Med 2015, 520–578.

[56] Yin SY, Peng AP, Huang LT, Wang YT, Lan CW, Yang NS. The Phytochemical Shikonin Stimulates Epithelial-Mesenchymal Transition (EMT) in Skin Wound Healing. Evidence-Based Complementary and Alternative Medicine 2016. http://dx.doi.org/10.1155/2013/262796

[57] Nikita G, Vivek P, Chhaya G. Wound-healing activity of an oligomer of alkannin/shikonin, isolated from root bark of Onosma echioides. Nat Prod Res 2015, 29(16), 1584–1588. doi: 10.1080/14786419.2014.986126

Navneet Kaur, Vikas Kumar, Parveen Rishi, Nirmala Sehrawat, Rahul Dilawari, Pawan Kumar, Neeraj Kumar Aggarwal

# 6 Phytomedicine: History, scope and future prospects

**Abstract:** Phytomedicine or herbal medicine has been playing a significant role in the treatment of diseases since time immemorial throughout human history. Medicinal plants are used as a medical resource in almost all cultures. Herbal medicines, which were the major source of remedy in the ancient system of medicine, have been employed in medical practices since antiquity. Herbal medication has gained importance in recent years because of its effectuality and cost effectiveness. This field is bringing forward new, safe and efficacious plant-based medicines. There has been increasing interest in the pharmaceutical industry to develop new medications from plants. During the past 50 years, great progress has been made in the chemical and biological techniques of analysis, which has transformed research in herbal medicine. Medicinal herbs/plants have an assured future since there are many plants around the globe, with most of them having not yet been studied in medical practice. Therefore, current and future studies on medical activities can be effective in treating diseases.

**Keywords:** Herbal medicines, Phytomedicine, Pharmaceutical, Biological Techniques

## 6.1 Introduction

Medicinal plants have long been used to treat many infectious diseases and human disorders for a large proportion of the population from ancient to modern times. The term phytomedicine was first coined by the French physician Henri Leclerc in 1913, who has published several essays on the use of medicinal plants (Phelps and Hassed). Phytomedicine, also known as herbal medicine, basically refers to the use of botanical raw material such as roots, seeds, leaves, bark, fruit bodies or flowers for medicinal and therapeutic purposes [1, 2]. The end of the last century witnessed a vigorous resurgence in the interest and utilization of medicinal plant products to sustain health. Contemporary phytomedicine has survived human testing throughout the ancient Chinese, Indian, Egyptian and Native American civilizations since time immemorial. A book written by Shennong Jing, *Herbal Classic of Shennong*, is one of the earliest known sources of folk knowledge of herbal medicines in China, dating back to 3000 B.C., which is a compilation of 365 plant-based medicaments [3]. Sprouted in the pristine land of India some 5000 years ago, Ayurveda is one of the

https://doi.org/10.1515/9783110563337-006

oldest healthcare systems in the world that believes in herbs as the solution to cure a number of health problems. The traditional knowledge of phytomedicines is inherited from ancestors and has evolved throughout the timeline of human civilization [2, 4]. Alternative systems of medicine became increasingly popular during the dawn of the scientific era, which greatly diminished the use of phytomedicine. However, recently there has been a shift in the global trend from synthetic to herbal medicine due to the revival of interest in its potential by the late 20th century. The emerging issues of cost effectiveness and potency of the drugs being faced by the conventional pharmaceutical industry are the major factors leading to the comeback of phytomedicine. According to the World Health Organization (WHO) reports in 2011, 80% of the people in majority of the developing countries still rely on traditional medicine for primary health care [5]. Due to the perpetual interest in herbal medicines, phytomedicine has come a long way with regard to the manufacturing techniques, quality control and steady accumulation of scientific evidence to elucidate the therapeutic mechanisms of action, efficacy and safety, hence revolutionizing their processing, development and usage (Phelps and Hassed). There are huge numbers of examples where whole plant or plant parts are used for the treatment of various diseases (Table 6.1). The increasing popularity and demand for herbal drugs can be attributed to the facts that natural plant products are nontoxic, have low side effects and are available at affordable prices compared to costly chemical drugs.

The beneficial medicinal effects of plant materials are typically due to the combinations of secondary products present in the plant. Secondary metabolites are substances produced by plants that have no proven role in growth, photosynthesis, reproduction or in other "primary" functions. These chemicals are extremely diverse; several thousands have been identified in many major categories. The medicinal actions of plants are unique and specific to particular plant species or groups, consistent with the concept that the combinations of secondary metabolites in a particular plant are often taxonomically distinct [6]. There are mainly three major groups of secondary metabolites, namely terpenes, phenolics, and N- and S-containing compounds. The renaissance of phytomedicine in the modern pharmaceutical industry necessitates studies in the fields of phytochemistry and pharmacognosy. Metabolic engineering could prove useful for modifying or enhancing the synthesis of valuable therapeutic agents present in medicinal plants.

## 6.2 History of phytomedicines

Primitive people used plants not only as food, but also for medicinal purposes, as has been noted by a number of sources providing ample historical evidence of human being's use of plants for medicinal purposes since ages [2, 7, 8]. The ancient Chinese, Indian, Egyptian and Native American civilizations all used to practice phytomedicine. The oldest written evidence of plant usage for medicinal drug

Table 6.1: List of some medicinally important plants and their phyto-medical applications.

| Common Name | Scientific Name | Parts Used | Medicinal Applications |
|---|---|---|---|
| **Amla** | *Emblica officinalis*<br>Fam: Euphorbiaceae | Fruit | Vitamin C, Cough, Diabetes, cold, Laxative, Hyperacidity |
| **Ashok** | *Saraca asoca*<br>Fam: Caesalpinanceae | Bark Flower | Menstrual Pain, Uterine, Disorder, Diabetes |
| **Aswagandha** | *Withania somnifera*<br>Fam: Solanaceae | Root, Leaves | Restorative Tonic, Stress, Nerve Disorder, Aphrodisiac |
| **Bael/Bilva** | *Aegle marmelos*<br>Fam: Rutaceae | Fruit, Bark | Diarrhea, Dysentery, Constipation |
| **Brahmi** | *Bacopa, Monnieri*<br>Fam: Scrophulariaceae | Whole Plant | Nervous, Memory Enhancer, Mental Disorder |
| **Guggul** | *Commiphora wightii*<br>Fam: Burseraceae | Gum Resin | Rheumatoid Arthritis, Paralysis, Laxative |
| **Guluchi/Giloe** | *Tinospora cordifolia*<br>Fam: Menispermaceae | Stem | Gout, Pile, General Debility, Fever, Jaundice |
| **Sandalwood** | *Santalum album*<br>Fam: Santalinaceae | Heart Wood, Oil | Skin disorder, Burning Sensation, Jaundice, Cough |
| **Sarpa Gandha** | *Ranwolfia serpentina*<br>Fam: Apocynaceae | Root | Hypertension, Insomnia |
| **Senna** | *Casia augustifolia*<br>Fam: Liliaceae | Dry Tubers | Rheumatism, Aphrodisiac |
| **Tulsi** | *Ocimum sanclum*<br>Fam: Lamiaceae | Leaves/Seed | Cough, Cold, Bronchitis |
| **Gritkumari** | *Aloe Verra*<br>Fam: Liliaceae | Leaves | Laxative, Wound Healing, Skin Burns & Care, Ulcer |
| **Bhringraj** | *Eclipta alba*<br>Fam: Compositae | Seed/Root/Whole | Anti-inflammatory, Digestive, Hair Tonic |
| **Neem** | *Azardirchata indica*<br>Fam: Mahaceae | Rhizome | Sedative, Analgesic, Epilepsy, Hypertensive |
| **Mandukparni/ Indian pennywort** | *Centella asiatica*<br>Fam: Umdelliferae | Whole Plant | Anti-inflammatory, Jaundice, Diuretic, Diarrhea |
| **Kaincha/Creeper Baidanka** | *Mucuna truriens*<br>Fam: Fabaceae | Root, Hair, Seed, Leaf | Nervous Disorder, Constipation, Nephropathy, Strangury, Dropsy |
| **Dalchini** | *Cinnamomum zeylanicum*<br>Fam: Lauraceae | Bark, Oil | Bronchitis, Asthma, Cardiac Disorder, Fever |

Table 6.1 (continued)

| Common Name | Scientific Name | Parts Used | Medicinal Applications |
|---|---|---|---|
| **Kurai** | *Holorheena antidysentrica* Fam: Apocyaceaceae | Bark, Seed | Scabies, Antipyretic, Amoebic Dysentery |
| **Kantakari/Akranti Perennil** | *Solanum xanthocarpum* Fam: Solanaceae | Whole Plant, Fruit, Seed | Diuretic, Anti-inflammatory, Appetizer, Stomach ache |
| **Mountain Daisy** | *Arnica Montana* Fam: Asteraceae | Flower | Anti-inflammatory, Analgesic, Antiseptic |
| **Mikvetch** | *Astragalus* Fam: Fabaceae | Roots, Rhizomes | Diuretic, Excessive Sweating, Ulcers, Diarrhea |
| **Black Cohosh** | *Cimicifuga racemosa* Fam: Ranunculaceae | Roots, Rhizome | Diuretic, Fatigue, Tuberculosis, Rheumatism |
| **Burdock** | *Arctium lappa* Fam: Asteraceae | Seeds, Leaves, Roots | Liver and Digestive Problems, Urinary Tract Infections, Ulcers, Eczema, Immune System Booster, Blood Purifier |
| **Bearberry** | *Arctostaphylosuva-ursi* Fam: Ericaceae | Leaves | Urinary Tract Infections and Inflammation |
| **Cat's Claw/ Hawk's Claw** | Uncaria tomentosa Fam: Rubiaceae | Bark, Root | Asthma, Bone Pain, Arthritis, Urinary Tract Infections, Ulcers, Rheumatism |
| **Kava** | *Piper Methysticum* Fam: Piperaceae | Rhizome, Roots | Diuretic, Arthritis, Asthma, Anxiety Disorders |
| **Ginkgo** | *Ginkgo biloba* Fam: Ginkgoaceae | Leaves, Seeds | Anticoagulant, Memory Booster |
| **Cloud Mushroom, Dancing Mushroom** | *Grifola frondosa* Fam: Meripilaceae | Root | Cancer and Autoimmune Disorders |
| **Lemongrass** | *Cymbopogon citrates* Fam: Poaceae | Grass | Nervous Disorders, Fevers, Arthritis, Flu, Gas, Pain, Reduce Anxiety |
| **Garlic** | *Allium sativum* (Liliaceae) | Bulb | Antihypertensive, Anti-hyperlipidemic, Platelet Aggregation Suppressant |
| **Fenugreek** | *Trigonella foenumgraecum* Fam: Fabaceae | Seeds | Respiratory Disorders, Cleanses the Stomach, Calms the Liver, Soothes Pancreas, Reduces Swelling |

preparation was found on a Sumerian clay tablet dated about 5,000 years ago by the ancient Sumerian of Mesopotamia. It described 12 recipes for drug preparations referring to over 250 different plants such as myrrh, opium laurel and caraway. These plants are still used all over the world for medicinal purposes. The Ebers Papyrus, written about 3,500 years ago, serves to illuminate the ancient herbal practices of the Egyptians. They used plants such as mandrake for pain relief and garlic for the treatment of heart and circulatory disorders. The Ebers Papyrus represents a collection of over 850 plant medicines, including garlic, juniper, cannabis, castor bean and aloe [2, 9]. A Chinese book "*Pen Ts'ao*", written by Emperor ShenNung (3000 B.C.), lists 365 medicinal plants and their uses, including camphor, ginseng, jimson weed, cinnamon bark and ephedrine, many of which are still in use (such as camphor, ginseng, jimson weed, cinnamon bark and ephedrine) [10, 9]. Ayurveda, the 5,000-year-old philosophy from India, includes the medicinal powers of culinary spices and herbs. Earliest Sanskrit writings such as the *Rig Veda* and *Atharva Veda* are some of the earliest available documents describing the medical knowledge that formed the basis of the Ayurveda system. Many herbs such as turmeric, nutmeg, pepper and clove used in Ayurveda were later described by ancient Indian herbalists such as Charaka and Sushruta during the 1st millennium BC. The *Sushruta Samhita,* attributed to Sushruta in the 6th century BC, describes the medicinal values of about 700 plant species [11, 12]. *The Hippocratic Corpus* serves as the first Greek herbal text that is associated with Hippocrates (400 B.C.), the famous Greek physician who used many herbal remedies in his practice. He is often referred to as the "Father of Modern Medicine". The early 19th century was a turning point in the knowledge and use of medicinal plants. It was the time when chemical analysis first became available and scientists began to extract and modify the active ingredients from plants. The formalization of phytomedicines led to a greater understanding of the specific actions of drugs on the body [13].

## 6.3 Techniques for the preparation of phytomedicines

Plant extracts are usually a mixture of various types of bioactive compounds or phytochemicals having different polarities and solubilities, which make their separation very challenging for their identification and characterization. A number of different separation techniques, such as thin layer chromatography (TLC), gel filtration chromatography and high-performance liquid chromatography (HPLC), can be used to obtain pure compounds, which can then be used to determine their structure and biological activity [3].

### 6.3.1 Chromatographic techniques

Chromatographic techniques have significantly contributed to the discovery of pharmaceutically important novel bioactive compounds from plants. Column chromatography allows the purification of compounds from a mixture using ion exchange, molecular sieves or adsorption. HPLC is a well-known analytical technique for the separation and determination of components of natural products [14]. Chemical separations using HPLC are based on different migration rates of certain compounds in a particular column and mobile phase. The extent or degree of separation is mostly determined by the choice of the stationary phase and the mobile phase. The bioactive phytochemical is often present only as a minor constituent in the crude extract, and the high resolving power of HPLC is ideally suited for the processing of such multicomponent samples on both the analytical and preparative scales [3].

### 6.3.2 Non-chromatographic techniques

Various non-chromatographic methods, such as monoclonal antibodies-based immunoassay (MAbs), Fourier-transform infrared spectroscopy (FTIR) and phytochemical screening assay, can also be utilized for easier identification of the bioactive phytochemicals. Immunoassays based on monoclonal antibodies are gaining importance in the analysis of bioactive compounds, especially for low-molecular-weight natural compounds. They are highly specific and sensitive for enzyme bioassays and for the qualitative as well as quantitative analytical techniques. FTIR is a highly valuable technique for the quantitative as well as qualitative analysis of organic and inorganic samples. It is an effective analytical instrument for the detection of functional groups and the characterization of covalent bonds in an unknown mixture of plant extract [15, 16]). In addition, FTIR spectra of pure compounds produce a profile in the form of a distinctive molecular fingerprint that can be used to screen samples for different components.

## 6.4 Secondary plant metabolites

The primary metabolites of plants produced through the fundamental metabolic pathways are macro- and micronutrients such as carbohydrate, fats, proteins and enzymes, nucleic acids and lipids, and essential oils. In contrast, secondary metabolites are organic compounds derived from primary metabolites that aid in the growth and development of plants but have no fundamental role in the maintenance of life processes in the plants [17]. The biologically active compounds present in plants as secondary metabolites, which elicit pharmacological or toxicological effects in humans and animals, are called **phytochemicals** [4]. The secondary metabolites produced by plants provide protection against plant pathogens and diseases and allow adaptation to their environments. They also exhibit a number of protective functions for human consumers. Many secondary metabolites are repellant or toxic to pathogens and herbivores. Secondary metabolites include representatives of all main classes of organic compounds: aliphatic, aromatic, hydroaromatic and heterocyclic. However, they largely fall into three classes of compounds: terpenoids, phenolics, and nitrogen-containing alkaloids and sulfur-containing compounds [18, 19]). Some of the major classes of pharmacologically active phytochemicals are listed in Table 6.2.

Table 6.2: List of medically important drugs/chemicals originating from plants with their clinical utilization.

| Drugs/Chemicals | Plant Source | Plant Part | Family | Action/Clinical Use |
|---|---|---|---|---|
| **Acetyldigoxin** | *Digitalis lanata* | Leaf | Plantaginaceae | Cardiotonic |
| **Aconitine** | *Aconitum napellus* | Root | Ranunculaceae | Analgesic, antipyretic, poisoning |
| **Ajmaline** | *Rauwolfia serpentina* | Root | Apocynaceae | Circulatory disorders |
| **Allicin** | *Allium sativum antimicrobial* | Bulb | Alliaceae | Arteriosclerosis, high cholesterol, heart disease, hypertension |
| **Berberine** | *Berberis vulgaris* | Root and bark | Berberidaceae | Antimicrobial, anti-inflammatory, antidiabetic, lowers cholesterol |
| **Caffeine** | *Coffea arabica* | Seed | Rubiaceae | CNS and metabolic stimulant |
| **Camphor** | *Cinnamomum camphora fungal infections* | Wood | Lauraceae | Cough, neurodermatitis |
| **Cocaine** | *Erythroxylum coca* | Leaf | Erythroxylaceae | Local anesthetic |
| **Colchicine** | *Colchicum autumnale* | Corm | Liliacae | Gout (also called gouty arthritis) |
| **Digitoxin, Digoxin (Digitalis or Digitalin)** | *Digitalis purpurea and other Digitalis species* | Leaf | Plantaginaceae | Cardiotonic |

Table 6.2 (continued)

| Drugs/Chemicals | Plant Source | Plant Part | Family | Action/Clinical Use |
|---|---|---|---|---|
| **Ephedrine** | *Ephedra sinica* and other members of *Ephedra* | Stem | Ephedraceae | Sympathomimetic, decongestant, hypotension |
| **Galantamine** | *Galanthus caucasicus* | Bulb and Flower | Amaryllidaceae | Alzheimer's disease, memory impairment |
| **Lycopene** | *Tomatoes,* other fruits and vegetables | | | Cancer prevention, heart disease, arteriosclerosis |
| **Menthol** | *Mentha species* | Leaf | Lamiaceae | Rubefacient |
| **Morphine, Codeine** | *Papaver somniferum* | Latex of immature seed pod | Papaveraceae | Analgesic, antitussive |
| **Nicotine** | *Nicotiana tabacum* | Leaf | Solanaceae | Insecticide |
| **Papain** | *Carica papaya* | Fruit | Caricaceae | Indigestion, skin and wound care, lowers blood pressure |
| **Quinine** | *Quinidine Cinchona* species | Bark | Rubiaceae | Antimalarial, antipyretic |
| **Salicin (Aspirin)** | *Salix alba* | Bark | Salicaceae | Analgesic, antipyretic, anti-inflammatory, prevents heart attack |
| **Taxol (Paclitaxel)** | *Taxus brevifolia* | Bark | Taxaceae | Antitumor/anticancer |
| **Vinblastine, Vincristine** | *Catharanthu sroseus* | Leaf | Apocynaceae | Anticancer, anti-leukemic, Hodgkin's disease |
| **Thymol**<br>**Strychnine** | *Thymus vulgaris*<br>*Strychnosnux-vomica* | Leaf<br>Seed | Lamiaceae<br>Loganiaceae | Antiseptic<br>CNS stimulant |
| **Diosgenin** | *Dioscorea species* | Tuber | Dioscoreaceae | Ovulatory, contraceptive |

## 6.5 Scope of phytomedicines in the health sector

All over the world in various developing countries, large numbers of people die daily of preventable or curable diseases because of the lack of even simple health care. Due to advancements in current research and financial investments, herbal medicines will, outwardly, continue to play an important part in improvement in the heath sector [20]. Herbal medicines prepared from plants are used for the maintenance of good health in most developing countries. In recent years, herbal medicines have gained importance due to their effectuality and cost effectiveness in treating various disorders/diseases. Discovery of drugs from medically important plants provides new and necessary leads against several medicine targets together with cancer, HIV or AIDS, Alzheimer's disease, malaria and pain, which are presently in late-phase clinical trials [21, 22]).

### 6.5.1 Herbal medicines in cancer prevention and treatment

Among the myriad health issues and diseases, development of chemotherapy against cancer is currently one of the most urgent concerns for public health. Cancer chemoprevention deals with the implication of natural and synthetic chemical compounds to reverse or suppress carcinogenesis and block the development of invasive cancers. [23] assessed the anticancerous potential of many natural sources, foods, supplements, crude drugs and Kampo medicines (traditional Japanese herbal substances) and observed that dietary non-nutrient compounds play important roles as chemo-preventive agents. Moreover, daily consumption of some particular food items has also been shown to have anticancer effects 24.

### 6.5.2 Therapeutic potential of phyto-molecules for obesity

Obesity is one of the major threats to human health in the 21st century. It has been declared a global epidemic by WHO as "New World Syndrome" [25, 26]. The unbalanced schedule of diet, such as high energy intake and low energy expenditure, leads to overweight and obesity [27]. Overeating, consumption of high dietary fat and less physical work collectively lead to obesity [28]. Dietary fat is also related to well-known diseases like diabetes, hypertension and cardiovascular diseases. Various pharmacological studies have reported that herbal medicines are promising options for the isolation, identification and characterization of bioactive compounds (flavonoids, saponins, alkaloids and anti-nutrients; kievitone and visoltricine) for developing effective anti-obesetic agents (Table 6.3). Natural products originating from plant sources as anti-obesity therapeutics have been reported as safer and effective alternatives for this rising epidemic [29, 30]).

Table 6.3: List of different plant species, phyto-molecules and their specific activities.

| Plant source | Phytochemical | Potential activity |
|---|---|---|
| ***Oryza sativa*** | Phenylboronic acid | Anti-obesetic agent |
| ***Salvia officinalis*** | Carnosic acid (diterpene) | Anti-obesetic agent |
| ***Platycodon grandiflorus*** | Platycodin D (saponin) | Anti-obesetic agent |
| ***Acanthopanax sessiliflorus*** | Sessiloside and chiisanoside (saponins) | Anti-obesetic agent |
| ***Dioscorea nipponica*** | Dioscin (saponin) | Anti-obesetic agent |
| ***Gardenia jasminoids*** | Crocin (terpene) | Anti-obesetic agent |
| ***Glycyrrhiz auralensis*** | Licochalcone A (polyphenol) | Anti-obesetic agent |
| ***Camellia sinensis*** | Proanthocyanidins and catechins | Pancreatic lipase inhibition activity |
| ***Cassia mimosoides*** | Proanthocyanidins | Pancreatic lipase inhibition activity |
| ***Malus domestica*** | Procyanidins | Pancreatic lipase inhibition activity |
| ***Eleutherococcus sessiliflorus, Aesculus turbinate, Panax ginseng, Dioscorea nipponica*** | Saponins | Pancreatic lipase inhibition activity |
| ***Arachis hypogaea, Nelumbo nucifera, Mangifera indica, Salacia reticulata*** | Polyphenols | Pancreatic lipase inhibition activity |
| ***Cassia nomame*** | Condensed tannins | Pancreatic lipase inhibition activity |
| ***Camellia sinensis*** | Theasaponins and polyphenols | Pancreatic lipase inhibition activity |
| ***Arnica montana*** | Phenols, flavonoids | Treatment of ostheoarthritis and rheumatoid arthritis |
| ***Equisetum arvense*** | Kynurenic acid | Treatment of ostheoarthritis and rheumatoid arthritis |
| ***Panax notoginseng*** | Saponins | Treatment of ostheoarthritis and rheumatoid arthritis |

Table 6.3 (continued)

| Plant source | Phytochemical | Potential activity |
|---|---|---|
| ***Sesamum indicum*** | Sesamin, sesamol, sesamolin | Treatment of ostheoarthritis and rheumatoid arthritis |
| ***Symphitum officinalis*** | Rosmarinic acids, glycopeptides, amino acids | Treatment of ostheoarthritis and rheumatoid arthritis |
| ***Zingiber officinalis*** | Gingerdione derivatives, 10-gingerol, 8,10-shogaol | Treatment of ostheoarthritis and rheumatoid arthritis |
| ***Whitania somnifera*** | Whitaferin A | Treatment of ostheoarthritis and rheumatoid arthritis |

These phyto-molecules reduce the absorption of fatty substances from food, diminish energy intake from diet, enhance energy consumption by metabolizing stored fat, decrease pre-adipocyte differentiation and proliferation, and increase the degradation of accumulated fat. Many enzymes related to fat digestion have been discovered and can be targeted for the development of therapeutics for obesity and related diseases [31]. Sukhdev and Singh [26] reviewed the therapeutic potential of phyto-molecules to manage obesity by retardation of nutrient absorption and digestion under various clinical trials. These molecules include valoneaic acid dilactone from Banaba, chestnut astringent skin extract and amylase inhibitor from white beans. In addition, acarbose against diabetes and cardiometabolic disorders, marine algae, fungi and some plants for anti-lipase activity are also helpful [32, 33, 34].

### 6.5.3 Role of phytomedicines in joint disorders

Chronic joint inflammatory disorders (osteoarthritis and rheumatoid arthritis) cause progressive histological alterations and produce disabling symptoms due to increase in inflammation and oxidative stress [35, 36]. The conventional medications (painkillers) used to prevent such disorders are commonly linked with serious, even fatal, side effects. Presently, medicinal plants are more promising alternatives to treat these disorders compared to conventional drugs. Many plant-derived natural products have been proven effective during *in vitro* studies or experiments on animal models; however, there is a lack of human clinical evidence in this regard. Dragos *et al.* [37] summarized the available scientific information on different medicinal plants having anti-arthritic activity tested on humans, listed in Table 6.3.

### 6.5.4 Phyto-molecules to cure oral diseases

Dental caries and periodontal diseases are major health issues in dental problems. This further affects the general quality of life. Moreover, poor oral health is associated with severe persisting conditions and systemic diseases. Earlier pieces of evidence showed harmful effects of some antibacterial drugs used in dentistry due to increase in resistance by bacteria toward antibiotics [38]. Presently, there is a need for an alternative possible prevention and treatment process that will be safer, more potent and cost effective to cure oral diseases. Natural phytochemicals isolated from plants used as traditional medicines are considered as good alternatives. It has been reported that plant extracts or phytochemicals have the capacity to inhibit the growth of oral diseases causing pathogens, lessen the development of biofilms and dental plaque, reduce the adhesion of microbes to surfaces and minimize the symptoms of oral diseases 39]. Many plant-derived compounds used traditionally as medicine have been recorded in pharmacopeias against oral microbial pathogens 40].

### 6.5.5 Phytomedicines to prevent metabolic diseases

The WHO encourages the addition of secondary metabolites of plants in life-saving drugs. Presently, poor diet pattern, stress, loss of physical activity and toxic substances in foods cause various disorders such as hypertension, diabetes mellitus, cardiovascular diseases, stroke, cancer and kidney diseases [41, 42]. Plant-based nutraceuticals and phytomedicines help prevent these metabolic diseases without any side effects. Identification and regular incorporation of these phytomedicines and nutraceuticals into the diet help increase the longevity of life. This also helps prevent synthetic-drug-induced physiological damage [42].

## 6.6 Challenges in the development of herbal medicine

The uses of traditional medicines are widespread, and plants represent a large source of natural chemicals that might serve as leads for the development of novel drugs. Despite the incredible history of traditional herbal medicines, their promotion faces a number of challenges related to regulatory status, assessment of safety and efficacy, quality control, safety monitoring and inadequate or poor knowledge about them [43]. The major limitations in the development of herbal medicine common to many countries are as follows.

1. **Quality issues**: Adulteration, misidentification of plant, faulty collection and preparation, and incorrect formulation process are the main problems reducing

the effectiveness of herbal preparation and can be considered as key factors affecting the quality and purity of herbal medicines.

2. **Standardization issues:** Standardization, poor quality-control procedure and lack of good manufacturing practices (GMPs) are the main hurdles in maintaining the quality of herbal drugs.
3. **Toxicity and adverse effects:** Herbal medicine can be efficacious as well as toxic as there is a specific dosage threshold for each herbal medicine. Side effects, contraindications and interactions with other drug, food and existing orthodox pharmaceuticals need to be monitor properly.
4. **Research and development:** Further research is needed to understand the mechanism of action and the pharmacokinetic phenomenon and for the improvement/creation of monographs. There is a lack of research on the development of high-yielding varieties.
5. **Climatic changes** that have adverse effects on the distribution and cultivation of medicinal plants also reduce their potential. These climatic changes are responsible for the loss of genetic resources in areas where they were found previously.

Indeed, the market and public demand has been so great that there is a great risk that many medicinal plants today face either extinction or loss of genetic diversity. To protect and preserve the knowledge of traditional medicines, there is a necessity to act firmly and rigorously. There should be proper ways to transfer this valuable information from generation to generation. Currently, this responsibility now lies on the shoulders of the regulatory bodies to monitor and ensure the quality of herbal products. They have to facilitate their development to clinical trial stages. The goals are closer to being achieved and can be achieved soon if the regulatory authorities work closely in association with research institutes, research clinics and centers, hospitals, industry and pharmacy faculties.

## 6.7 Future prospects

Phytomedicine is considered both preventive and therapeutic in nature. The efficacy of some herbal products is beyond doubt; some of the most recent examples include *Artemisia annua* (i.e. artemesinin: wormwood derivative used for anticancer and antimalarial properties) [44–48], *Taxus brevifolia* (i.e. taxols: pacific yew derivative that exhibits anticancer activity) [49] and *Silymarin marianum* (i.e. silymarin: seeds of milk thistle effective in treating liver diseases) [50, 3]. With the advent of sophisticated techniques such as HPLC, gas chromatography (GC) and GC-mass spectrometry (MS), it is possible to identify as well as quantify the test substance (51). Medicinal herbs/plants have an assured future since there are many plants around the globe, with most of them not yet being studied in medical practice. Therefore, current and future studies on medical activities can prove effective in treating diseases.

# References


[1] Falodun A (2010) Herbal medicine in Africa-distribution, standardization and prospects.*Res J Phytochem* 4(3): 154–161.

[2] Chikezie PC, Ojiako OA (2015) Herbal Medicine: Yesterday, Today and Tomorrow. *Altern Integr Med* 4: 195.

[3] Sasidharan S, Chen Y, Saravanan D, Sundram KM, Yoga Latha L (2011). Extraction, isolation and characterization of bioactive compounds from plants' extracts. *Afr J Tradit Complement Altern Med.*; 8(1):1–10. Sheng-Ji P (2001). Ethnobotanical approaches of traditional medicine studies: some experience from Asia. *Pharmaceutical Biology*; 39 (l): 74–79.

[4] Mohamed I, Shuid A, Borhanuddin B, Fozi N (2012). The application of phytomedicine in modern drug development. *The Internet Journal of Herbal and Plant Medicine*; 1(2).

[5] Robinson MR & Zhang X. The world medicines situation 2011 (Traditional medicines: global situation, issues and challenges). Geneva: World Health Organization; 2011.

[6] Wink M (1999) Introduction: biochemistry, role and biotechnology of secondary products. in Biochemistry of Secondary Product Metabolism. ed Wink M (CRC Press, Boca Raton, FL), pp 1–16.

[7] Kong JM, Goh NK, Chia LS, Chia TF (2003) Recent advances in traditional plant drugs and orchids. *Acta Pharmacol* Sin 24: 7–21.

[8] Philipeon JD (2003) Phytochemistry and medicinal plants. Phytother 56:237–243

[9] Biljana Bauer Petrovska (2012). Historical review of medicinal plants' usage *Pharmacogn Rev.*; 6(11): 1–5.

[10] Wiart C (2006). Etnopharmacology of medicinal plants. New Jersey: Humana Press; pp. 1–50.

[11] Sumner J. (2000).The natural history of medicinal plants. Timber Press, Portland, USA. (http://www.cabdirect.org/abstracts/20003017295.html)

[12] Girish Dwivedi, Shridhar Dwivedi (2007). History of Medicine: Sushruta – the Clinician – Teacher par Excellence (PDF). National Informatics Centre. Archived from the original (PDF) on 2008-10-10. Retrieved 2008-10-08.

[13] Dervendzi V (1992). Contemporary treatment with medicinal plants. Skopje: Tabernakul; pp. 5–43.

[14] Cannell, R. J. P. (1998). Natural Products Isolation. New Jersey: Human Press Inc; 1998. pp. 165–208.

[15] Eberhardt, T.L., Li, X., Shupe, T.F. and Hse, C.Y. (2007). Chinese Tallow Tree (Sapium Sebiferum) utilization: Characterization of extractives and cell-wall chemistry. *Wood Fiber Sci.* 39: 319–324.

[16] Hazra, K. M., Roy R. N., Sen S. K. and Laska, S. (2007). Isolation of antibacterial pentahydroxy flavones from the seeds of *Mimusops elengi* Linn. *Afr. J. Biotechnol.* 6 (12): 1446–1449

[17] Rungsung W, Ratha KK, Dutta S, Dixit AK, Hazra J (2015). Secondary metabolites of plants in drugs discovery. *World Journal of Pharmaceutical Research*; 4 (7): 604–617

[18] Edreva A, Velikova V, Tsonev T, Dagnon S, Gurel A, Aktaş L, Gesheva E (2008). Stressprotective role of secondary metabolites: diversity of functions and mechanisms. *Gen. Appl. Plant Physiology*; 34(1-2): 67–78.

[19] Veberic R (2010). Bioactive compounds in fruit plants; 4–5.

[20] Hoareau, L. and E.J. DaSilva, 1999. Medicinal plants: A re-emerging health aid. Electron. *J. Biotechnol.*, 2: 56–70.

[21] Sumner, Judith (2000). The Natural History of Medicinal Plants. Timber Press. 17. ISBN 0-88192-483-0.

[22] Mamedov N and Craker L (2008). Endemic species of Artemisia L. (Asteraceae) from Caspian Sea region as alternative sources for malaria treatment. Proceedings of the First International Medicinal and Aromatic Plants Conference on Culinary Herbs, Antalya, Turkey. Acta Horticulturae. 826, 277–280.
[23] Yasukawa K (2012). Medicinal and Edible Plants as Cancer Preventive Agents, Drug Discovery Research in: *Pharmacognosy*. ISBN: 978-953-51-0213-7.
[24] Banning M (2005). The carcinogenic and protective effects of food. *British Journal of Nursing*;14(20):1070–1074
[25] Caballero B. (2007) The global epidemic of obesity: An overview, Epidemiol Rev, 29, 1–5.
[26] Sukhdev Singh and Kanwar Shamsher Singh, (2013) Therapeutic Role of Phytomedicines on Obesity: Importance of Herbal Pancreatic lipase Inhibitors. International Research Journal of Medical Sciences, 1, 15–26.
[27] Astrup A. (2001) The role of dietary fat in the prevention and treatment of obesity. Efficacy and safety of low-fat diets. Int J Obes 25, 46–50.
[28] Schrauwen P and Westerterp KR. (2000) The role of high-fat diets and physical activity in the regulation of body weight. Br J Nutr 84, 417–427.
[29] Soetan KO and Oyewole OE. (2009)The need for adequate processing to reduce the anti-nutritional factors in animal feeds: A review. Afr J Food Sci 3, 223–232.
[30] Muhammad A, Dangoggo SM, Tsafe AI, Itodo AU, Atiku FA. (2011) Proximate minerals and anti-nutritional factors of Gardenia aqualla fruit pulp. Pak J Nutr 10, 577–581.
[31] Armand M. (2007) Lipases and lipolysis in the human digestive tract: where do we stand? Curr Opin Clin Nutr Metab Care 10, 156–164.
[32] Tsujita T, Takaku T and Suzuki T. (2008) Chestnut astringent skin extract, an alpha-amylase inhibitor, retards carbohydrate absorption in rats and humans. J Nutr Sci Vitaminol 54, 82–88.
[33] Slanc P, Doljak B, Kreft S, Lunder M, Janes D and Strukelj B. (2009) Screening of selected food and medicinal plant extracts for pancreatic lipase inhibition. Phytother Res 23, 874–877.
[34] Yamagishi S, Matsui T, Ueda S, Fukami K and Okuda S. (2009) Clinical utility of acarbose, an alpha-glucosidase inhibitor in cardiometabolic disorders, Curr Drug Metab, 10, 159–163.
[35] Yang, C.L.H.; Or, T.C.T.; Ho, M.H.K.; Lau, A.S.Y. (2013). Scientific Basis of Botanical Medicine as Alternative Remedies for Rheumatoid Arthritis. Clin. Rev. Allergy Immunol. 44, 284–300.
[36] Hsu, D.-Z.; Chu, P.-Y.; Jou, I.-M. (2016). Daily sesame oil supplement attenuates joint pain by inhibiting muscular oxidative stress in osteoarthritis rat model. J. Nutr. Biochem. 29, 36–40.
[37] Dorin Dragos, Marilena Gilca, Laura Gaman, Adelina Vlad, Liviu Iosif, Irina Stoian and Olivera Lupescu. (2017) Phytomedicine in Joint Disorders. Nutrients, 9, 70-88. doi:10.3390/nu9010070
[38] Petersen, P. E., D. Bourgeois, H. Ogawa, S. Estupinan-Day, and C. Ndiaye, (2005) "The global burden of oral diseases and risks to oral health," Bulletin of the World Health Organization, 83, 661–669.
[39] Enzo A. Palombo, (2011). Traditional medicinal Plant Extracts and Natural Products with Activity against Oral Bacteria: Potential Application in the Prevention and Treatment of Oral Diseases, Evidence-Based Complementary and Alternative Medicine, Article ID 680354, 15 pages, doi:10.1093/ecam/nep067
[40] D. Kalemba and A. Kunicka, (2003) "Antibacterial and antifungal properties of essential oils," Current Medicinal Chemistry, 10, 813–829.
[41] Anderson GF, Chu E. (2007) Expanding priorities: confronting chronic disease in countries with low income. N Engl J Med., 356: 209–211.
[42] Girija K. and Radha R. (2013) Role of phytomedicines and neutraceuticals in the prevention of chronic diseases, International Journal of Pharmaceutical sciences and Research, 4, 3363–3368.

[43] Thillaivanan S, Samraj K (2014). Challenges, Constraints and Opportunities in Herbal Medicines – A Review. *International Journal of Herbal Medicine*; 2 (1): 21–24.

[44] Ferreira JF, *Luthria* DL, *Sasaki* T, *Heyerick* A, *(2010)*. Flavonoids from Artemisia annua L. as Antioxidants and Their Potential Synergism with Artemisinin against Malaria and Cancer. *Molecules*. 15 (5): 3135–3170.

[45] Efferth T, Benakis A, Romero MR, Tomicic M, Rauh R, Steinbach D, Häfer R, Stamminger T, Oesch F, Kaina B, Marschall M (2004). Enhancement of cytotoxicity of artemisinins toward cancer cells by ferrous iron. *Free Radical Biology and Medicine*. **37**(7): 998–1009.

[46] Nakase, I; Gallis, Byron; Takatani-Nakase, T; Oh, S; Lacoste, E; Singh, NP.; Goodlett, DR.; Tanaka, S; Futaki, S (2009). Transferrin receptor-dependent cytotoxicity of artemisinin–transferrin conjugates on prostate cancer cells and induction of apoptosis. *Cancer Letters*. 274 (2): 290–298.

[47] Lai HC, Singh NP, Sasaki T *(2013). Development of artemisinin compounds for cancer treatment. Invest New Drugs*. 31 (1): 230–46.

[48] Abolaji AO, Eteng MU, Ebong PE, Brisibe EA, Dar A, Kabir N, Choudhary MI (2013). A safety assessment of the antimalarial herb Artemisia annua during pregnancy in Wistar rats. *Phytotherapy research: PTR*. **27**(5): 647–54.

[49] Kingston AD (1993) "Taxol, an exciting anticancer drug from *Taxus brevifolia*: An overview" In: Kinghorn AD, Balandrin MF (eds) *Human Medicinal Agents from Plants* (Washington, DC: American Chemical Society):138–48.

[50] Phelps K, Hassed C. General Practice: The Integrative Approach by Kerryn Phelps, Craig Hassed (chapter 4 principles of phytomedicines).

[51] Bazala V (1943). The historical development of medicine in the Croatian lands. Zagreb: Croation publishing bibliographic institute. pp. 9–20.

[52] Yasukawa K. In: Medicinal and Edible Plants as Cancer Preventive Agents, Drug Discovery Research in: Pharmacognosy. Prof Omboon Vallisuta., editor. InTech; 2012. [March 14, 2013]. ISBN: 978-953-51-0213-7

Diwakar Aggarwal, Anil Kumar, Amit Kumar

# 7 Plant tissue culture for commercial propagation of economically important plants

**Abstract:** The exploration of plant tissue culture implies aseptic development of explants *in vitro* in culture media comprising supplements and plant hormones and their hatching under controlled physical conditions (light, temperature and humidity). Tissue culture has been a subject of enthusiasm for quite a while, and lately it has turned into a valuable tool for agribusiness and medicines. Recently, tissue culture has developed as a science with a huge potential for large-scale plant production, particularly in horticulture and forestry. Quality planting materials of numerous species have been produced on a mass scale through tissue culture. The present chapter reviews the *in vitro* propagation studies of some commercially important plant species like *Eucalyptus, Populus* and *Aloe.*

**Keywords:** Micro-propagation, Agroforestry, Medicinal Plants, Horticultural and Ornamental Plants

## Abbreviations

| | |
|---|---|
| BA | 6-benzyl adenine |
| 2iP | 2-isopentenyl adenine |
| IAA | indole-3-acetic acid |
| IBA | indole-3-butyric acid |
| Kn | 6-furfurylaminopurine |
| PGRs | plant growth regulators |
| MS | Murashige and Skoog (1962) medium |
| NAA | α-naphthalene acetic acid |
| GA3 | gibberellic acid |
| TDZ | thidiazuron (1-phenyl-3 (1, 2, 3-thiadiazol-5-yl) urea |
| BAP | 6-benzylaminopurine |
| WPM | woody plant medium Lloyd and McCown (1981) |
| 2,4-D | 2,4-dichlorophenoxyacetic acid |

https://doi.org/10.1515/9783110563337-007

## 7.1 Introduction

Plant tissue culture is a technique to cultivate parts of plants (cell, tissue and organs) and develop them on artificial media under aseptic conditions in a controlled condition so parts cultivated *in vitro* can develop into complete plants [1]. The technique of tissue culture was initiated to demonstrate the hypothesis of "totipotency" as depicted by Schleiden and Schwann in 1838, which expresses that the plant cell as the smallest unit can develop and flourish if kept under controlled environmental conditions [2]. For the first time, German plant physiologist Gottlieb Haberlandt (1854–1945) attempted the *in vitro* cultivation of cells in 1902 [3] and is thus credited with the title of father of plant tissue culture (Table 7.1). Developing throughout the years, plant tissue culture now includes a number of *in vitro* systems, techniques and procedures utilized for the proliferation of plants under *in vitro* conditions. Tissue culture has been used to create genetic variability in plants to produce some changes and to build the quantity of elite germplasm accessible to the plant breeder. Tissue culture techniques, in combination with recombinant DNA technology, have been effectively used to fuse particular attributes through gene transfer [4]. *In vitro* procedures through protoplast, anther, microspore and ovule cultures have been utilized to make new varieties in the reproducing lines regularly through haploid production. Somaclonal variants with novel traits have also been created with cell culture techniques. Tissue culture methods have also been utilized for the generation of disease-free plants. These methods have become very important for successful plant genetic engineering [5].

Tissue culture has been utilized to produce plants on a mass scale by utilizing the technique of micro-propagation (Figure 7.1). Many plants with the same hereditary characters can be propagated using this technique. Subsequently, this technique has become an option for the vegetative proliferation of plants [6]. Micro-propagation provides true-to-type plants of a selected genotype utilizing three *in vitro* culture methods: (1) the axillary-bud multiplication approach, (2) organogenesis and (3) somatic embryogenesis. The axillary-bud proliferation approach is further subdivided into shoot and meristem culture, where the explants are taken from the growing shoot tips or the terminal/axillary buds usually with the stem segment attached to it. Plant propagation through organogenesis is achieved by two modes: organogenesis through callus formation and the emergence of adventitious organs directly from the explant; the latter method is preferred for mass multiplication of plants. Embryogenesis is achieved either directly or indirectly. In the direct method, embryos directly originate from explants without any callus, whereas in the indirect method, first the callus is formed from the explant and embryos develop later from the callus [7]. Although the axillary-bud proliferation approach produces minimal number of plantlets as

Table 7.1: Landmark discoveries in the field of plant tissue culture.

| Year | Discovery (Worker) |
|---|---|
| 1902 | First attempt to culture isolated plant cells *in vitro* (Gottlieb Haberlandt, called father of plant tissue culture) |
| 1904<br>1922 | Embryo culture of selected crucifers (Hannig)<br>*In vitro* culture of root tips (Robbins) |
| 1934 | Successful culture of tomato roots (White) |
| 1939 | Successful establishment of growing callus culture (Gautheret, Nobecourt and White) |
| 1944 | *In vitro* adventitious shot formation in tobacco (Skoog) |
| 1952 | Use for meristem culture to obtain virus-free plants (Morel and Martin) |
| 1955 | Discovery of kinetin (Miller *et al.*) |
| 1957 | Discovery of regulation of organ formation with the cytokinin and auxin ratio (Skoog and Miller) |
| 1960 | Enzymatic degradation of cell walls to obtain a large number of protoplasts (Cocking) |
| 1962 | Development of the Murashige and Skoog Nutritional medium (Murashige and Skoog) |
| 1964 | Production of haploid embryos (Guha and Maheshwari) |
| 1970 | First achievement of protoplast fusion (Power *et al.*) |
| 1976 | Shoot initiation from cryopreserved shoot apices of carnation (Seibert) |
| 1977 | Successful integration of the Ti plasmid DNA from *Agrobacterium tumefaciens* in plants (Chiton *et al.*) |
| 1981 | Introduction of the term somoclonal variation (Larkin and Scowcrort) |
| 1984 | Genetic transformation of tobacco with *Agrobacterium tumefaciens* (De Block *et al.*, Horch *et al.*) |
| 1990 | RAPD technique developed (Willams *et al.*) |

the quantity of shoots created is controlled by the quantity of the axillary buds cultured, it remains the most broadly utilized strategy in commercial plant tissue culture as it creates the most consistent plantlets. Unfortunately, somatic embryogenesis, which has the capability of creating the highest number of plantlets, has not been used to its potential thus far [8]. Although seedling-raised plants are widely used as explants for micro-propagation, they have the drawback of having an obscure genetic background [9]. Utilizing mature plants as a starting material has the issues of high contamination rate and inconsistent multiplication rates, coupled with poor rooting and germination response. Yet, the utilization of mature plants as the starting material is attractive because its genetic background is known [10, 11]. Endangered, threatened and rare species have effectively been proliferated by micro-propagation owing to its high multiplication rate and needing a minimal number of starting material for establishing aseptic cultures. With time, micro-propagation has become the most common method of propagation of plants that are generally difficult to propagate by seeds or other vegetative means [12]. Micro-propagation protocols are available for most crop species, although continued optimization is still required for many crops, especially cereals and woody plants [13].

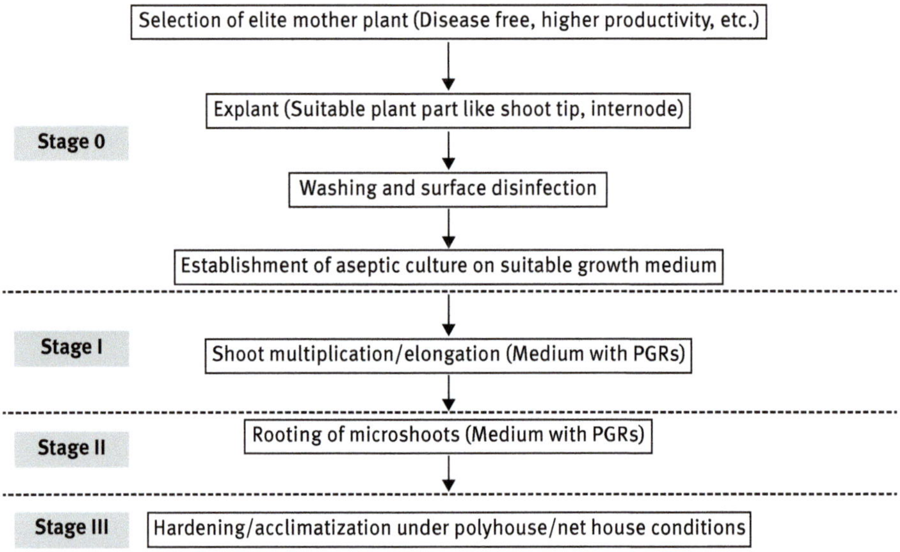


Figure 7.1: General outline of micropropagtaion protocol with respective stages.

Micro-propagation offers several advantages over other propagation methods, e.g. fast multiplication rate, generation of disease-free plants, production round the year, germplasm conservation and encouraging their easy trade through exchange of material. Still, the success of plant tissue culture is hindered by several factors likewise [14]. Issues of vitrification and contamination can cause great loss in a tissue-culture lab [15]. Genetic variation leading to change in ploidy level has been accounted for loss of desirable economic traits in the tissue-cultured plants [16]. High input cost coupled with costly and specialized prerequisites, along with trained labor, significantly increases the production cost. Yet, micro-propagation remains the most common method of propagation for the economic-level *in vitro* production of plants. This chapter discusses the different methodologies used for the *in vitro* propagation of plants with economic significance.

# 7.2 Forestry/Agroforestry plants

Because of fast depletion of the forest cover and exhaustion of elite genetic stocks, solid endeavors must be made to develop new strategies for the mass proliferation and generation of short rotation trees with a quick turnover of biomass and initiation of genetic variability. Other strategies used for vegetative propagation like grafting, budding and layering have failed to produce the desirable results [17]. Hence, it is envisaged that *in vitro* propagation techniques are capable of addressing this issue. *In vitro* propagation is of specific significance to tree species that have long age cycles confounded with the issue of heterozygosity. Tissue culture strategies have been used for the mass-scale proliferation of many important crops, and a number of commercial research centers have been set up in numerous parts of the world for the large-scale propagation of quality planting materials. However, its optimized used for forest tree species production has just begun as of late [18]. Strategies for tissue culture are currently accessible for mass multiplication of *Eucalyptus* and *Populus* trees. In addition, there are different reports where tissue culture techniques have been used for the propagation of *Dalbergia sissoo, D. latifolia, Albizia lebbeck*, sandal, rubber and teak [19, 20].

## 7.2.1 Eucalyptus

*Eucalyptus* (family Myrtaceae), an enduring and evergreen species of plants, is among the most widely planted hardwood trees on the planet [21]. Native to Australia, *Eucalyptus* has more than 700 species [22], and these fast-growing trees were rapidly embraced for plantation forestry programs by different countries, including India [23]. The immense financial interest in *Eucalyptus* species is because of the flexibility of their wood [24]. Its timber has applications in a wide range of purposes, e.g. pulp and paper creation, electric posts, charcoal, timber and furniture. Lately, there has been increased interest for utilizing its wood as an inexhaustible, financially viable option to petroleum derivatives as biofuels. Short-duration plantings of *Eucalyptus* will probably be an important source of wood feedstock [21] as they can deliver a huge amount of cellulose-rich biomass that could be used for bioenergy and biopolymer generation [25]. In view of their outstanding diversity and adaptability, the growth area under its cultivation is evaluated at around 20 million ha [26], spread out across more than 37 nations and representing 16% of the overall forest area [27]. India has the biggest zone under *Eucalyptus* cultivation (8 million ha), followed by Brazil (3 million ha) [28]. *Eucalyptus* plantations contribute 25% of the aggregate wood consumed in the developed countries and utilized primarily for obtaining fiber by the pulp and paper industry [29].

Although most of the *Eucalyptus* species can be propagated vegetatively utilizing conventional stem-cutting methods, they have several limitations, e.g. poor establishment of stem cuttings, graft incompatibility problems [30, 31], poor accessibility of propagules and affected by season. Therefore, micro-propagation is progressively becoming the preferred method for the establishment of clonal plantations. The significance of *in vitro* propagation of *Eucalyptus* is clear from the amount of work carried out on *Eucalyptus* utilizing different micro-propagation procedures [32, 09, 33, 34, 35, 15, 36, 37]. The raising of micro-propagated plants from various types of *Eucalyptus* through different techniques has been successfully carried out [32, 38, 39, 40, 37]. This opened up the possibility for substantial clonal propagation of elite clones. Additionally, higher yield was obtained from micro-propagated plants compared with seedling-raised plants [41]. Therefore, the potential effects of micro-propagation on forest productivity, wood-based industry and worldwide timber supplies are presently very much acknowledged [42].

For initiation of aseptic cultures, different plant parts, such as cotyledons, hypocotyls and leaf pieces taken from *in vitro*-raised seedlings, have been utilized as explants [35, 15, 36, 37]. Other than taking explants from *in vitro*-developed plantlets, explants from coppiced shoots [32, 43], scion shoots [44, 45], epicormic shoots [46] and young shoots from actively growing trees [32, 09; 47, 48] have additionally been utilized as starting materials. Time and season of collection of explants are important factors in the establishment of aseptic cultures [15, 49]. After initiation of aseptic cultures, shoot multiplication and shoot lengthening were accomplished on different medium combinations. Das and Mitra [50] reported an increase in the number of shoots on MS medium supplemented with 0.5 μM NAA and 4.4 μM BA utilizing shoot tips as explants taken after coppicing of mature trees of *E. tereticornis*. Sharma and Ramamurthy [15] observed shoot initiation and increase in number of shoots on MS medium supplemented with 4.4 μM each of BA and NAA. Lowering the concentration of PGRs led to shoot elongation. *Eucalyptus* hybrid (*Eucalyptus erythronema* var *erythronemax, Eucalyptus stricklandii* cv. “Urrbrae Gem”) was successfully micro-propagated by Glocke *et al.* [51] through an improved axillary shoot initiation method. Shoot multiplication was higher on WPM and QL [52] media supplemented with BA, NAA and $GA_3$ when contrasted with MS, B5, AP (53) and TK [54] media supplemented with the same concentrations of PGRs. Aggarwal *et al.* [32] investigated various factors impacting the micro-propagation of an elite clone of *E. tereticornis* (Figure 7.2A to 2E). Among the various cytokinins tried, 6-BA turned out to be the best cytokinin for increasing the number of shoots and elongation. Auxins are generally utilized for the induction of roots in microshoots. Similarly, for establishing *Eucalyptus* microshoots, various auxins have been reported to be broadly utilized [32, 15, 31, 55]. Of the different auxins tried, viability of IBA for the induction of roots in microshoots of different plant species has been depicted in detail by Kato [56] and was demonstrated to be more powerful than different auxins like NAA and IAA for induction. The concentration and method of use of IBA change with various plant species [57, 56]. Furthermore, reducing the salt concentration was found to be useful for establishing roots in microshoots in *Eucalyptus* [32, 31]. A major problem in commercial-scale use of plant tissue culture is the high mortality experienced by tissue-culture-raised plants following laboratory to land transfer, primarily because of the contrasting differences between the *in vitro* and *ex vitro* conditions. It was additionally revealed that plants propagated under *in vitro* conditions have poor photosynthetic assembly [58]. Various strategies have been devised for acclimatization of tissue-culture-raised plants for fruitful field establishment of *Eucalyptus* [32, 59, 60]. Aggarwal *et al.* [32] observed that plants produced under photosynthetic active radiation (PAR) light showed higher survival rates following their transfer to soil, which commensurate with higher chlorophyll contents. Furthermore, they revealed that inoculation of plantlets with bacterial isolates during acclimatization was observed to be useful for the survival of plants. Grijashanker [59] revealed the utilization of various soil mixtures for effective acclimatization of the

**Figure 7.2: Micropropagation of elite plants of *E. tereticornis* and *P. deltoides* A.** Shoot multiplication (E.T) on MS medium supplemented with 2.5 µM BA and 0.5 µM NAA **B.** Shoot elongation (E.T) on MS medium supplemented with 0.1 µM BA and 0.5 µM NAA **C.** Microshoots rooted (E.T) on one fourth-strength MS medium supplemented with 5.0 µM IBA **D.** Acclimatized plantlets (E.T) under greenhouse conditions **E.** 2 year old tissue culture raised plants (E.T) growing in the field **F.** Shoot multiplication (P.D) on MS medium supplemented with 2.5 µM BA **G.** Shoot elongation (P.D) on MS medium supplemented with 0.5 µM BA and 0.1 µM NAA **H.** Microshoots rooted (P.D) on MS medium supplemented with 5.0 µM IBA **I.** Acclimatized plantlets (P.D) under greenhouse conditions. (E.T.- *Eucalyptus tereticornis*, P.D.- *Populus deltoides*).

micro-propagated plantlets of *E. camaldulensis*. To further improve the survival rate, covering of micro-propagated plantlets with a polythene cover was found to be beneficial.

### 7.2.2 Populus

The genus *Populus* L. (poplars, cottonwoods and aspens) consists of around 30 species [61] that are generally found distributed on the northern side of the equator. They are the deciduous trees belonging to the family Salicaceae (willow family). The most common species belonging to the genus are *Populus deltoides, P. trichocarpa, P. tomesto, P. nigra and P. angustifolia*. They vary in leaf size, shape and shade of the bark. Although they have a short life expectancy, they are one of the quickest growing trees, growing up to 5–10 feet every year [62]. *Populus* has found multipurpose uses in assembling plywood, fiber board, almirahs, false roofs, sports merchandise and counterfeit appendages. Its fibers have additionally demonstrated an amazing property for making different types of papers, like fine paper, pressing paper and newsprint. Its attributes like straight clean bole, leaflessness in winter, soil-enhancing properties, similarity to rural yields and high monetary returns make this tree species very adaptable and the most perfect for planting on rural fields, organic product plantations, steers sheds, cultivate streets and in homesteads [63, 64]. In addition, poplars are extremely attractive for biofuel generation because of their quick development and ability to deliver a huge amount of biomass in a brief period of time, and hence they are considered as important factors for woody bioenergy programs. [65, 66]. Additional attributes like small genome measure, quick development to adolescence, productive sexual proliferation and the accessibility of broad hereditary maps have made the *Populus* tree species very popular among woody plants [67, 68].

The tissue of *Populus* displays a high level of versatility, similar to that of tobacco in the herbaceous species, and hence they have been the subject of much research in plant tissue culture. *Populus* species is the main timberland tree species for which *in vitro* procedures have been characterized to date for recovery from protoplasts of leaf mesophyll tissue, sporophytic calli, gametophytic (anther) calli, intemodes, root sections, axillary buds and apical meristems. Although many commercially important *Populus* species have been propagated through *in vitro* propagation methods, *P. deltoides* remains the first choice as it is the most widely cultivated species of the genus *Populus* throughout the world. In India, *P. deltoides* clones have demonstrated superb performance in Uttar Pradesh, Punjab, Haryana and Maharashtra states since its introduction [69].

Recovery from plant tissues *in vitro* is to a great extent subject to explant decision, medium structure and control of the physical environment [70, 71]. Murashige [72] recognized a few factors that ought to be considered in explant choice including the organ selected as the tissue source, the physiologic and ontogenetic age of the organ and the general nature of the mother plant. Subsequently, in *Populus* likewise, different explants like *in vitro* leaf, internode, nodes, petiole and root explants were utilized for the development of *in vitro* culture. Thakur *et al.* (69) observed the highest frequency of shoot recovery (74.75%) from petiole explants on MS medium supplemented with 0.50 mg/L BAP and 0.20 mg/L IAA.

Cavusoglu *et al.* [73] reported the use of node, internode and leaf explants for the induction of somatic embryogenesis on MS medium supplemented with 0.05 mg/L BA and 5 mg/L 2, 4-D for establishing aseptic cultures of *P. deltoides*. Various explants responded differently, and the maximum response was observed from internodes, followed by leaves and nodes. Noel *et al.* (74) reported the use of stem cuttings having at least one axillary bud harvested from coppiced shoots as explants for the establishment of aseptic cultures in six European poplar cultivars of economic interest – four *Populus × interamericana* and two *Populus × euramericana*. Chaturvedi *et al.* [75] reported the use of nodal segments for the establishment of aseptic cultures in the case of *P. deltoides*, whereas Kang *et al.* [76] used shoot tips as an explant in the case of *P. trichocarpa*. Maheshwari and Kovalchuk [77] utilized stem internodes and axillary-bud explants (estimate ~0.5–1.0 cm) for the initiation of *in vitro* cultures. These were washed completely under running water for 30 min. They were then cleaned for 15 min in a 10% (v/v) blanch solution and then further disinfected with 70% ethanol for 1.5 min. Finally, the explants were washed four times with sterile water before inoculation. Song *et al.* [78] reported the segments from the 1st to the 10th stem internodes, with the third and fourth leaves and the midribs from these leaves, as explants. Aggarwal *et al.* [79] used petioles as explants in the case of *P. ciliate*. From the above-mentioned studies, it is clear that type and age of explants have profound effects on the establishment of *in vitro* cultures, and one should be careful while selecting explants for economically important trees like *Populus*.

Jiang *et al.* [80] reported the utilization of BA and NAA for shoot organogenesis in *Populus × euramericana* "Neva". Maximum regeneration frequency along with the number of shoots/explant from shoot tips (96.7%, 9.8) and leaves (90.0%, 8.7) was achieved on the half-quality MS medium supplemented with 0.5 mg/L BA and 0.1 mg/L NAA. The ideal medium for *in vitro* establishing of roots was half-quality MS medium containing 1 mg/L IBA maximum rooting frequency (93.3%) along with the number of roots/explant (8.2). For acclimatization, *in vitro*-obtained plantlets were transferred to polyhouse containing vermiculite and peat (1: 1). Aggarwal *et al.* [79] utilized thidiazuron (TDZ) alone and mixed with adenine and NAA for the recovery of shoots in *P. cilate*. High shoot organogenesis frequency was (86 %) recorded on MS medium supplemented with 0.004 mg/L TDZ and 79.7 mg/L adenine. Shoots were further elongated on MS medium supplemented with 0.5 mg/L BAP mixed with 0.2 mg/L IAA and 0.3 mg/L $GA_3$. *In vitro* rooting was accomplished on MS medium supplemented with 0.10 mg/L IBA alone. *In vitro*-raised plantlets were acclimatized for 2 months in a sterile blend of sand and soil. Thakur *et al.* [69] developed an effective and quick procedure for the *in vitro* recovery of *P. deltoides* clone G48. The maximum regeneration frequency (74.75%) was acquired on MS medium supplemented with 0.50 mg/L BAP and

0.20 mg/L IAA. Shoots regenerated in this way begin turning darker and necrotic following 10–15 days in culture. To address the issue of phenolic leach-out, shoot buds were adjusted to MS medium containing 0.50 mg/L BAP, 0.20 mg/L IAA, 15 mg/L AdS, 0.1% PVP, 100 mg/L casein hydrolysate, 50 mg/L glutamine, 250 mg/L $(NH_4)_2SO_4$ and 0.5% agar. Shoot elongation was accomplished on a similar medium. IBA at 0.10 mg/L was observed to be the best for *in vitro* rooting of microshoots. The *in vitro*-recovered plantlets were effectively acclimatized and set up in nursery conditions. Kang *et al.* [76] developed a profoundly productive micro-propagation procedure for *Populus trichocarpa* (Torr. and Gray) genotype "Nisqually-1". They examined the impacts of basal medium and cytokinin fixation on plant development. MS, woody plant medium (WPM) [81] or Driver and Kuniyuki walnut (DKW) [82] medium was supplemented with 0.1% MS vitamins, 100 mg/L myo-inositol, 3% (w/v) sucrose and 0.8% (w/v) agar. Additionally, these basal media were supplemented with three groupings of BA, i.e. 0.0, 2.2 and 4.4 μM. Furthermore, the impacts of gelling agent and activated charcoal on micro-propagation were studied. It was found that shoots developed on MS medium without BA had higher survival rates than those developed in MS medium with 2.2 or 4.4 μM BA, or in WPM or DKW medium with or without BA. Gelrite was observed to be the preferred gelling specialist over agar. Addition of activated charcoal into the medium enhanced the multiplication rate, and plantlets were successfully acclimatized under greenhouse conditions.

## 7.3 Medicinal plants

Conventional medications play a fundamental role in human progress by curing different diseases. As indicated by the World Health Organization (WHO), up to 80% of the general populations of the world have been dependent on herbs as a customary solution to cure different ailments since the start of human civilization [83]. Natural products act as a leading source for the discovery of new drugs [84]. Because of this reestablished enthusiasm for natural drugs, at least 40% of the pharmaceutical organizations have begun utilizing formulations based on plants. This has prompted increased request of normal crude material, leading to the exhaustion of some plant species. Although plant tissue culture is not an elective technique for the mass-scale propagation of medicinal plants [85], it can be utilized for the biosynthesis and biotransformation of biologically active compounds, for the stockpiling of plant cells and organs, and for genetic engineering of plants.

### 7.3.1 Catharanthus roseus

*Catharanthus roseus* (Linn.) G. Don., commonly known as Madagascar periwinkle, belonging to the family Apocynaceae is a medicinal plant of enormous commercial value owing to the presence of 120 types of bioactive terpenoid indole alkaloids

(TIAs) in its leaves and roots [86] and is known worldwide for its anticancerous property. It is used to treat diseases like diabetes, menstrual problems, hypertension and, most importantly, cancer in a number of countries [87]. The increasing demand of *C. roseus* in recent years has led to its widespread use, leading to the scarcity of quality planting material [88]. To meet the needs of important alkaloids present in different parts of *C. roseus*, researchers have started using *in vitro* propagation methods for their cultivation. The *in vitro*-proliferated plants with medicinal value provide a uniform, sound and good plant material for biochemical portrayal and the identification of active constituents [89].

Endo *et al.* [90], Hirata *et al.* [91], Junaid *et al.* [92] and Swamberg and Dai [93] aseptically propagated plants under *in vitro* conditions from seeds of *C. roseus* on MS medium supplemented with different concentrations of BA. In most of the cases, seeds were disinfected with 70% ethanol for 2 min and 20% blanch solution for 15 min, followed by five washes with sterile distilled water. After surface disinfection, they were set on MS medium in a petri dish for germination. The developed seedlings were propagated until they reached a length of 2–4 cm. Different parts (nodal stem, leaf, root and hypocotyl) were later utilized as explants for developing aseptic cultures. Verma and Mathur [94] utilized leaves without petioles from 4–6-month-old glasshouse-grown plants of known varieties of *C. roseus* as explants. Later, Kumar *et al.* [95] utilized nodal portions with apical buds (around 5–6 cm from the top) as explants from 1–2-year-old plants of *C. roseus*. In both the cases, nodal explants were first washed in running water with a couple of drops of Tween-20 and then treated with 1,000 ppm cefotaxime and 1,000 ppm bavistin for 10 min to remove dust particles as well as parasitic and bacterial spores. Finally, the surface was disinfected with 0.1% mercuric chloride for 15 min followed by flushing the explants five times with sterile (autoclaved) distilled water aseptically Nodal sections of around 0.5–0.8 cm were arranged aseptically and were inoculated vertically on MS medium consisting of mixes of BAP and NAA for the development of shoots from axillary buds.

Various researchers [96, 92, 93, 94, 97, 95] studied the impacts of various blends of auxins, cytokinins and other vital components, like light intensity, genotype and so on, on the emergence of microshoots of *C. roseus* under *in vitro* conditions. The authors showed that the hormones BA and NAA added to the MS medium affected the emergence of shoots at various concentrations although 2,4-D was found to be effective for callus formation. Yuan *et al.* (96) additionally showed that light intensity of 550–700 Lux was observed to be advantageous to the development of shoots on MS medium supplemented with 2 mg/L of 6-BA and 0–1.0 mg/L of NAA.

Junaid *et al.* [92] reported an efficient somatic embryogenesis method for *C. roseus*. Friable embryonic callus was induced from hypocotyl explant on MS medium supplemented with various auxins like NAA and 2, 4-D (1.0 mg/L). However, only NAA (1.0 mg/L) produced somatic embryos in cultures. The rate of embryo development increased when the same medium was added with BAP. Embryos were successfully converted into plantlets on MS medium supplemented with

BAP (0.5 mg/L) in combination with $GA_3$ (1.0 mg/L). Furthermore, carbon sources and concentrations were found to have a marked influence on the maturation process. Fructose and maltose (3–6 %) were found to be better carbon sources for the maturation process compared with other carbon sources.

Swanberg and Dai [93] examined the impacts of genotype, explant and PGR on plant recovery under *in vitro* conditions. Callus formation was observed from both leaf and multipurpose explants of all cultivars when cultured on WPM with PGRs. The best frequency of callus formation (100%) was observed on the media containing both BA and NAA. The cultivars demonstrated restricted recovery from leaf tissues, and the recovery rate was significantly lower compared with that from multipurpose tissues. Shoot recovery was observed to be PGR and genotype dependent. Maximum regeneration frequency (73.3%) was accomplished when internodes were inoculated on WPM containing 5 μM BA and 5 μM NAA, although 56.7% of inter-nodal explants created shoots on WPM with 20 μM BA and 10 μM NAA. Carbenicillin and cefotaxime had little impact on the regeneration frequency. Microshoots thus obtained were rooted easily. Maximum rooting frequency (100%) was observed on MS medium supplemented with 5.0 μM NAA, and the roots developed were shorter and thicker than typical roots. About 60% of the rooted plants survived in the nursery and blossomed within 2–3 months.

Verma and Mathur [94] revealed the procedure for high recurrence of shoot buds from the leaves of *C. roseus*. First, they pretreated the leaf explants for 60 minutes in a cell protoplast washing medium containing 13% (w/v) mannitol taken after their plating on a half-quality MS medium supplemented with 7.0 mg/L BA and 3.0 mg/L NAA. The treatment brought about the emergence of shoot buds in over 75% of the explants. Histological studies showed that shoot buds arise from hypodermal tissues around the mid-rib. The rooting in the recovered shoots was observed on MS media supplemented with 3.0 mg/L indole-3-butyric acid (IBA), and the established plants could be effectively settled in soil with a 70% survival rate.

The impact of plant hormones on callus formation utilizing distinctive explants, from various explants, was studied by Singh *et al.* [97]. An aggregate of 19 shoot recovery media was utilized by the authors, resulting in callus formation with the exception of MS Basal. Furthermore, three medium combinations were used to observe the regeneration potential of explants. For shoot multiplication, the MS basal medium supplemented with BAP (1.5 mg/L) with NAA (1.0 mg/L) was found to be satisfactory, followed by the MS medium supplemented with BAP (3.0 mg/L) with NAA (4.0 mg/L). Reducing the potency of MS media to half quality in the mixture of IBA (2.5 mg/L) and NAA (0.5 mg/L) gave the best rooting response.

A proficient micro-propagation procedure for *C. roseus* was reported by Kumar *et al.* [95]. Experiments were conducted to optimize suitable media for *in vitro* shoot multiplication and root induction. Of the various media tested for *in vitro* shoot multiplication, the MS medium supplemented with 1 mg/L of BA and

0.2 mg/L of NAA indicated better reaction as far as the growth of shoots from axillary buds is concerned. The microshoots rooted effectively on half quality of MS medium having 1 mg/L IBA and 0.25% charcoal. RAPD investigation showed that the micro-propagated plants were hereditary like the mother plants and also were hereditarily uniform too.

Aseptic cultures of *C. roseus* can be specifically initiated, with high frequencies, from seeds, seedlings, shoot meristems and axillary-bud explants taken from mature plants. *In vitro*-raised plant material gives a greater frequency of *C. roseus* plant material for the isolation of alkaloids having medicinal properties. The outcomes discussed in this section demonstrate that biotechnological strategies for *in vitro* proliferation provided new resources of exceptionally profitable crude material.

### 7.3.2 Aloe vera

*Aloe vera* syn. barbadensis Mill. (Liliaceae) is a high-value medicinal plant that flourishes in varied climatic conditions. As indicated by Leuang [98], *Aloe* species originated in South Africa and it comprises roughly 500 species found mostly in South Africa, but now distributed throughout the world. *Aloe vera* was first described by Carl Linnaeus in 1753 as *Aloe perfoliata* var. *vera*. It was named after the Greek word *Alloeh*, which means a shining bitter substance. It is a stable succulent xerophyte growing up to 1.5 m tall, with a solid root. The leaves are whitish green on the both sides and have barbed-teeth-like edges. It propagates vegetatively, and its propagation rate is low [99]. Most parts of the plant are utilized by the drug and cosmetic industries [100, 101]. It is generally used in the Ayurveda system to cure skin diseases. *Aloe* leaves contain a gel-like substance that is extremely effective in curing wounds, burns and many other diseases [102]. It is also reported to have anticancer, cell reinforcement, mitigating, immunomodulatory, hepato-protective, antiulcer and antidiabetic properties [103]. The plant is reported to contain cancer prevention agents like vitamins A, C and F, niacin, riboflavin and folic acid and can be utilized as a purgative, anti-helminthic and uterine stimulant [104]. Indiscriminate uses of wild stock have led to its threatened status, requiring scientific endeavors for its preservation and mass multiplication. Because of its low seed feasibility, slow germination rate, constrained accessibility of high-quality crude material and moderate vegetative development, tissue culture can be used for the preservation of these high-value medicinal plants [105, 106, 107]. *In vitro* investigations of different *Aloe* species by utilizing different explants have been accounted for by different researchers [108, 109, 100, 110, 111, 112].

For the initiation of aseptic cultures, explants were generally gathered from field-developed plants, which are generally infected by different microorganisms. Other than surface contaminants, there are numerous interior infectants, otherwise called endophytes, that can regrow even after years in culture. Normal cleaning agents like sodium or calcium hypochlorite (5–10%), ethyl alcohol (50–95%) and mercuric chloride (0.01–0.1%) are utilized to destroy the surface contaminants by washing in proper solution for 5–15 min depending on the nature of tissue and explant utilized followed by a few flushes in sterile water. Similarly, in the case of *Aloe* also, various explants like meristems, shoot tip, shoot crown and leaf explants were used for establishing aseptic cultures using standard tissue culture protocols. After the surface disinfection and inoculation of explants, it took 1–4 weeks for the shoots to emerge under *in vitro* conditions [100, 110,111, 112].

After initiation of aseptic cultures, various media combinations were utilized for the *in vitro* propagation and *in vitro* rooting of *Aloe*. However, the medium depicted by Murashige and Skoog [113] (MS medium) is the most frequently utilized, often with minor changes. Molsaghi *et al.* [114] utilized the MS medium supplemented with diverse plant hormones like 1 mg/IAA mixed with 4 mg/L BAP, 1 mg/L Kin with 0.1 mg/L IAA, 0.5 mg/NAA with 2 mg/L BAP and 0.2 mg/L IAA in a blend of 0.8 mg/L BAP for shoot establishment and shoot multiplication. The development of shoots occurred in 1 week. Maximum shoot regeneration frequency was found in a mix of 4 mg/L BAP with 1 mg/L IAA. Shoot elongation with the greatest length was acquired on MS medium supplemented with 1 mg/Kin mixed with 0.1 mg/L IAA. Elongated microshoots were rooted on MS medium supplemented with 1 mg/L IAA. These established plantlets with developing roots (5–7 cm long) were transferred to plastic pots containing garden soil, fertilizer and sand in the ratio of 1:1:1, separately, with 100% survival rate.

Sahoo and Rout [115] conducted experiments from leaf explants of *Aloe barbadensis* Mill on MS medium supplemented with 2.0 mg/L 6-BA or Kn mixed with 0.25–0.5 mg/L NAA for shoot regeneration. Maximum shoot regeneration frequency of shoot buds was observed on MS medium supplemented with 2.0 mg/L BA with 0.5 mg/L NAA additionally supplemented with 40 mg/L of adenine sulfate. Supplementation of 0.25–0.50 mg/L gibberellic acid into the medium prompts shoot lengthening. Microshoots were rooted on half-quality MS medium supplemented with 0.25–1.0 mg/L NAA or IBA with 2 % (w/v) sucrose. Maximum frequency of rooting was obtained on medium having 0.5 mg/L NAA with 3% (w/v) sucrose. *In vitro* plantlets were acclimatized in the nursery conditions with 80% survival rate and effectively settled natural conditions. RAPD and ISSR markers were utilized for setting up the clonal constancy of the *in vitro*-propagated plantlets with the mother plant.

Aggarwal and Barna [100] had reported a micro-propagation protocol for an elite plant of *Aloe vera* syn *A. barbadensis* through the enhanced axillary branching method. After establishment of aseptic cultures on MS medium containing 1 mg/L BA and 0.2 mg/L IBA, shoot multiplication was also achieved on the same medium. Addition of citric acid at 10 mg/L improved the shoot multiplication. It was observed that 100% microshoots produced rooted plantlets within 15 days of culture on a hormone-free agar medium. The plants were successfully transferred in the soil and were morphologically similar to their mother plants (Figure 7.3 D–F).

Figure 7.3: **Micropropagation of *Catharanthus roseus* and *Aloe vera* A.** Shoot multiplication (C.R) on MS medium supplemented with 10.0 µM BA and 5.0 µM NAA **B.** Microshoots rooted (C.R) on Half-strength MS medium supplemented with 5.0 µM IBA **C.** Acclimatized plantlets (C.R) under greenhouse conditions **D.** Shoot multiplication (A.V) on MS medium supplemented with 2.5 µM BA in combination with 0.2 µM IBA **E.** Microshoots rooted (A.V) on MS basal medium **D.** Acclimatized plantlets (A.V) under greenhouse conditions. (C.R- *Catharanthus roseus,* A.V- *Aloe vera* ).

Abrie and Van Staden [116] reported a quick *in vitro* proliferation method for the highly endangered species *Aloe polyphylla*. Plantlets gathered from *in vitro*-developed seeds were cultured on MS medium containing BA alone or a mix of BA and NAA. After the initial issues of phenol leaching, the explants quickly shaped into axillary and adventitious buds. Maximal shoot development was observed on MS medium supplemented with 1.0 mg/L BA. Microshoots were effectively established on MS medium supplemented with 0.5 mg/L of IBA. Rooted plantlets were acclimatized to nursery conditions.

Although significant work on the *in vitro* proliferation of aloe has been carried out, there is still much left to be studied. Significant progress has been made on different methods of *in vitro* propagation ranging over different species, but the results have not been quite satisfactory in the cases of haploid production, anther culture and protoplast culture with special reference to callus culture and somatic embryogenesis. Very little progress has been made in the fields of biochemical and molecular markers in the case of aloe. Considerably more endeavors are required that will help researchers grow more productive proliferation methodologies.

## 7.4 Horticultural crops

Horticultural crops mainly involve the multiplication of ornamental and fruit plant species. Most horticultural crop species have traditionally been propagated using vegetative techniques so that desirable traits are preserved. However, plants produced through vegetative modes of propagation showed some problems like crowding, less adaptability to environmental conditions, with sometimes diseases being transferred to the offshoots, loss of vigor, inability to introduce new characters and difficulty in removing undesirable characters [117]. Therefore, to address these problems, *in vitro* culture methods are preferred for the propagation of economically important ornamentals. There have been three essential objectives of utilizing a part of *in vitro* culture strategies with ornamentals: (1) generation of disease-free planting material, (2) the fast propagation of a large number of genetically similar plants and (3) the introduction of new and improved genotypes. Most tissue culture programs have focused on the following factors: (1) mass proliferation of attractive lines and rootstocks, (2) production of disease-free plants, (3) quick micro-propagation of attractive lines from rearing projects, (4) safeguarding of germplasm and (5) generation of haploids for plant reproduction.

Fruit crops that have received the maximum consideration with respect to tissue culture are banana, strawberries, apples, peaches, grapes and berries [118]. A review of the literature demonstrates that most researchers are utilizing the MS medium with slight modifications. In certain European nations, the larger part of all commercially propagated strawberries is created from *in vitro*-inferred plantlets. While commercial labs in the United States are propagating chrysanthemum and dianthus *in vitro*, and various commercial labs are associated with the mass multiplication of ornamentals through tissue culture systems, in India, the micro-propagation industry is providing significant help to the Indian agribusiness regarding four crops: fruits, ornamentals, spices and forestry/plantation [119]. India is the second biggest producer of fruit plants after China. Mango, banana, citrus, guava, grape, pineapple and apple are the significant plants propagated in India. In addition, different plants like papaya, ber, *phalsa*, sapota, annona, jackfruit and pomegranate in the tropical and subtropical locations and peach, apricot, pear, almond, walnut and strawberry in the temperate locations are propagated through tissue culture. Banana is the biggest tissue culture food crop in India (Table 7.2). Propagation in fruit plants through seeds is fundamentally performed to raise rootstocks, which are required for grafted plants. A portion of the fruit plants like papaya is engendered through seeds, yet dormancy, poor germination, low seed practicality and unfavorable environmental conditions are real constraining elements [120]. TC papaya plants are currently utilized for the extraction and handling of papain. Lately, production of an extensive variety of fancy plants by tissue culture has become an attractive business practice. The rapid strides made in the field of micro-propagation of fancy plants have been made possible by the efforts of various devoted specialists. In spite of the fact that it is difficult to

Table 7.2: Commercially micro-propagated plants in India.

| Plant Category | Plants |
|---|---|
| Fruits | Banana, Pineapple, Strawberry, Papaya, etc. |
| Spices | Turmeric, Ginger, Vanilla, Large Cardamom,Small Cardamom |
| Medicinal plants | Aloe Vera, Geranium, Stevia, Patchouli, Neem |
| Cash crops | Sugarcane, Potato |
| Ornamentals | Gerbera, Carnation, Anthurium, Lily, Syngonium, Cymbidium |
| Woody plants | Teak, Bamboo, Eucalyptus, Populus |
| Biofuel | Jatropha, Pongamia |

discuss these achievements in this concise outline, a portion of the more critical commitments are essential, from a chronicled point of view. Among ornamentals, *in vitro*-raised orchids and gerberas have achieved business significance. TC rose plants are utilized as pot plants.

Much more progress will be made in the coming decades to further develop plant propagation and other interesting and attractive plant attributes. Enhanced parallel production of therapeutic, aromatic and pharmaceutical offers extraordinary guarantee [121, 122]. Much progress has been made in this field as confirmed by reports from a few labs utilizing bioreactor frameworks to produce secondary metabolites [123, 124]. The use of bioreactors for micro-propagation, as noted by Ziv, will undoubtedly be further studied and refined. Regardless of whether micro-propagation is the "influx without bounds" or just an intense instrument for the propagation of plants, it is certain that the fate of micro-propagation and all of agriculture is surely brilliant and energizing.

## References

[1] Thorpe TA. Frontiers of plant tissue culture. Calgary: University of Calgary Printing Services; 1978, 556.
[2] Bouga JM, von Aderkas P. *In vitro* culture of tress. Kluwer Acad Publ. Dordrecht 1992, 231.
[3] Haberlandt G. Kulturversuche mit isoliserten Pflanzenzellen. Sitzungsber. Akadder Wiss. Wien, Math.-Naturwiss. 1902, K1, 111:69–92.
[4] Brown DCW, Thorpe TA. Crop improvement through tissue culture. World J Microb Biot 1995, 11, 409–415.
[5] Mantell SH, Matthews JA, McKee RA. Principles of plant biotechnology. Blackwell Scientific, Boston, 1985, 130–157.
[6] George EF, Hall MA, Klerk GJD. Plant Tissue Culture Procedure – Background. In: George EF, Hall MA, Klerk GJD (eds) Plant Propagation by Tissue Culture. Springer, Dordrecht, 2008.
[7] Thorpe TA. History of plant tissue culture. Molecular Biotechnology 2007, 37, 169–180.
[8] Thorpe T A. The current status of plant tissue culture. In S. S. Bhojwani (Ed.), Plant tissue culture: Applications and limitations Elsevier, Amsterdam 1990, 1–33.

[9] Aggarwal D, Kumar A, Reddy MS. Shoot Organogenesis from elite plants of *Eucalyptus tereticornis*. Plant Cell Tiss Org Cult 2010,102, 45–52.
[10] Jones NB, Van Staden. Micropropagation of *Eucalyptus*. In: Bajaj YPS (ed) High-tech and micropropagation V, vol 39. Springer-Verlag, Berlin Heidelberg New York (Biotechnology in agriculture and forestry) 1997, 286–329.
[11] Jain SM. An update on overall recent progress on somatic embryogenesis in forest trees. In: Suzuki K, Ishii K, Sakurai S, Sasaki S (eds) Plantation technology in tropical forest science, Springer, Tokyo 2006, 113–122.
[12] Yasodha R, Sumathi R, Gurumurthi K. Micropropagation for quality production in plantation forestry. Indian J Biotech 2004, 3, 159–170.
[13] Debergh P, Zimmerman RH. Micropropagation: Technology and Application. Springer Netherlands 1991, doi 10.1007/978-94-009-2075-0.
[14] Wang PJ, Charles A. Micropropagation through Meristem Culture. In: Bajaj Y.P.S. (eds) High-Tech and Micropropagation I. Biotechnology in Agriculture and Forestry, Springer, Berlin, Heidelberg 1991, 17.
[15] Sharma SK, Ramamurthy V. Micropropagation of 4-year-old elite *Eucalyptus tereticornis* trees. Plant Cell Rep 2000 19, 511–518.
[16] Rani V, Raina SN. Genetic analysis of enhanced-axillary branching derived *Eucalyptus tereticornis* Smith and *E. camaldulensis* Dehn. Plants. Plant Cell Rep 1998, 17, 236–242.
[17] Liu X, Pijut PM. Plant regeneration from *in vitro* leaves of mature black cherry (*Prunus serotina*). Plant Cell Tiss Org Cult 2008 94:113–123.
[18] Aggarwal D, Reddy MS, Kumar A. Biotechnological Approaches for the Improvement of Eucalyptus.. Plant Tissue Culture: Propagation, Conservation and Crop Improvement. doi 10.1007/978-981-10-1917-3_1, 2016, 219–244.
[19] Vibha JB, Shekhawat NS, Mehandru P, Dinesh R. Rapid multiplication of *Dalbergia sissoo* Roxb.: a timber yielding tree legume through axillary shoot proliferation and ex vitro rooting. Physiol Mol Biol Plants 2014 20(1), 81–87.
[20] Yasodha R, Sumathi R, Gurumurthi K. Improved micropropogations methods for Teak. J Trop For Sci 2005, 17(1), 63–75.
[21] Myburg AA, Grattapaglia D, Tuskan GA, Hellsten U, Hayes RD, Grimwood J. et al. (2014) The genome of *Eucalyptus grandis*. Nature 2014, 509, 356–362.
[22] Brooker MIH. A new classification of the genus *Eucalyptus* L'Her(Myrtaceae).. Aust Syst Bot 2000, 13, 79–148.
[23] Doughty RW. The *Eucalyptus*- A natural and commercial history of the gum tree. The Johns Hopkins University Press, Baltimore and London 2000, ISBN 0-8018-6231-0.
[24] Endo T, Goodbody A, Misawa M. Alkaloid production in root and shoot cultures of *Catharathus roseus*. Plan Med 1987, 53, 479–482.
[25] Mizrachi E, Mansfield SD, Myburg AA. Cellulose factories: advancing bioenergy production from forest trees. New Phytol 2012, 194 (1), 54–62.
[26] GIT Forestry. Cultivated eucalyptus global map. http://www.git-forestry.com, 2008.
[27] FAO. The global outlook for future wood supply from forest plantations by C. Brown. Working paper GFPOS/WP/03, Rome 2000, 129.
[28] Junghans DT, Alfenas AC, Brommonschenkel SH, Oda S, Mello EJ, Grattapaglia D. Resistance to rust (*Puccinia psidii* Winter) in *Eucalyptus*: mode of inheritance and mapping of a major effect locus with RAPD markers. Theor Appl Genet 2003, 108, 175–180.
[29] Zhou H. Isolation and functional genetic analysis of *Eucalyptus* wood formation genes. Dissertation, Department of Genetics, University of Pretoria, Pretoria, South Africa, 2005.
[30] Vengadesan G, Pijut PM. *In vitro* propagation of northern red oak (*Quercus rubra* L.). In Vitro Cell Dev-Pl, 2009, 45, 474–482.

[31] Bennett IJ, McDavid D, McComb JA. The influence of ammonium nitrate, pH and indole butyric acid on root induction and survival in soil of micropropagated *Eucalyptus globules*. Biol Plant 1994, 47, 355–360.
[32] Aggarwal D, Kumar A, Sharma J, Reddy MS. Factors Effecting micro propagation and acclimatization of an elite clone of *Eucalyptus tereticornis*. In Vitro Cell Dev-Pl 2012, 48, 521–529.
[33] Dibax R, Deschamps C, Bespalhok Filho JC, Vieira E, Molinari C, Campos D, Quoirin M, Organogenesis and *Agrobacterium tumefaciens*-mediated transformation of *Eucalyptus saligna* with P5CSgene. Biol Plantarum 2010, 54(1), 6–12.
[34] Pinto G, Park YS, Silva, Neves L, Araujo C, Santos C. Factors affecting maintenance, proliferation and germination of secondary somatic embryos of *Eucalyptus globulus* Labill. Plant Cell Tiss Org Cult 2008, 95, 69–78.
[35] Prakash MG, Gurumurthi K. Somatic embryogenesis and plant regeneration in *Eucalyptus tereticornis* Sm. Curr Sci 2005, 88, 1311–1316.
[36] Termignoni RR, Wang PJ, Hu CY. Somatic embryo induction in *Eucalyptus dunnii*. Plant Cell Tiss Org Cult 1996, 45, 129–132.
[37] Subbaiah MM, Minocha SC. Shoot regeneration from stem and leaf callus of *Eucalyptus tereticornis*. Plant Cell Rep 1990, 9, 370–373.
[38] Dibax R, Eisfeld CL, Cuquel FL, Koehler H, Quoirin M. Plant regeneration from cotyledonary explants of *Eucalyptus camaldulensis*. Scientia Agricola (Piracicaba, Brazil) 2005, 62, 406–412.
[39] Mullins KV, Llewellyn DJ, Hartney VJ, Strauss S, Dennis ES. Regeneration and transformation of *Eucalyptus camaldulensis*. Plant Cell Rep 1997, 16, 787–791.
[40] Chang SH, Donald DGM, Jacobs G. Micropropagation of *Eucalyptus radiata* ssp. *radiata* using explants from mature and coppice material. South African Forestry J 1992 162:43–47.
[41] Khuspe SS, Gupta PK, Kulkarni DK, Mehta V, Mascarenhas AF. Increased biomass production by *Eucalyptus. Can J Forest Res* 1987, 17, 1361–1363.
[42] Yanchuk AD. The role and implications of biotechnological tools in forestry. Unasylva 2001, 52, 53–61.
[43] Burger DW. *In vitro* micropropagation of *Eucalyptus sideroxylon*. HortScience 1987, 22, 496–497.
[44] Franclet A, Boulay M. Micropropagation of frost resistant eucalypt clones. Aust J Forest Res 1982, 13, 83–89.
[45] Goncalves AN. Reversion to juvenility and cloning of *Eucalyptus urophylla* S.T. Blake in cell and tissue culture systems. IUFRO Symposium and Workshop on Genetic Improvement and Productivity of Fast Growing Tree Species. S Lo Paulo, Brazil 1980, 786–787.
[46] Ikemori YK. Epicormic shoots from the branches of *Eucalyptus grandis* as an explant source for in vitro culture. Commonwealth Foreign Rev 1987, 66, 351–355.
[47] Rao KS. *In vitro* meristem cloning of *Eucalyptus tereticornis* Sm. Plant Cell Rep 1988, 7, 546–549.
[48] De Fossard RA, Barker PA, Boume RA. The organ culture of nodes of four species of *Eucalyptus*. Acta Hortic 1977, 78, 157–l 65.
[49] Grewal S, Ahuja A, Atal CK. *In vitro Proliferation of shoot apices of Eucalyptus* citridora Hook. Indian J Exp Biol *1980*, 18, 775–777.
[50] Das T, Mitra GC. Micropropagation of *Eucalyptus tereticornis* Smith. Plant Cell Tiss Org Cult 1990, 22, 95–103.
[51] Glocke P, Delaporte K, Collins G, Sedgley M. Micropropagation of juvenile tissue of *Eucalyptus erythronema x Eucalyptus Stricklandii* cv. 'urrbrae gem'. In Vitro Cell Dev-Pl 2006, 42, 139–143.
[52] Quoirin, M, Lepoivre P. Improved medium for *in vitro* culture of *Prunus* sp. Acta Hort 1977, 78, 437–442.
[53] Almehdli AA, Parfitt DE. *In vitro*, propagation of peach: Propagation of 'Lovell' and 'Nemguard' peach rootstocks. Fruit Varieties J 1986, 40, 12–17.

[54] Tabachnik L, Kester DE. Shoot culture for almond and almond-peach hybrid clones in vitro. HortScience 1977, 12, 545–547.
[55] Gupta PK, Mehta UJ, Mascarenhas AF. A tissue culture method for rapid clonal propagation of mature trees of *Eucalyptus torelliana* and *Eucalyptus camaldulensis*. Plant Cell Rep 1983, 2, 296–299.
[56] Kato M. Regeneration of plantlets from tea stem callus. *JPN J Breed* 1985, 35, 317–322.
[57] Jha T, Sen SK. Micropropagation of an elite Darjeeling tea clone. Plant Cell Rep 1992, 11, 101–104.
[58] Kozai T. Acclimatization of micropropagated plants. In: Bajaj YPS (eds) High- Tech and Micro propagation V (Biotechnology in Agriculture and Forestry), Springer- Verlang, Berlin Heidelberg, 1991, 17, 127–141.
[59] Girijashanker V. *In Vitro* regeneration of *Eucalyptus camaldulensis*. Physiol Mol Biol Plants 2012, 18(1), 79–87.
[60] Machado R, Andrade D, Cid B, Penchel RM, Brasileiro M. *Agrobacterium* strain specificity and shooty tumour formation in *Eucalyptus* (*Eucalyptus grandis × E. urophylla*). Plant Cell Rep 1997, 16, 299–303.
[61] Park W, Baba K, Furuta Y, Iida I, Sameshima K, Arai M, Hayashi T. Enhancement of growth and cellulose accumulation by overexpression of xyloglucanase in poplar. FEBS Letters 2004, 564, 183–187.
[62] Stanturf JA, van Oosten, C, Netzer, DA, Coleman M D, Portwood CJ. Ecology and silviculture of poplar plantations, in *Poplar* Culture in North America, eds Dickmann DI, Isebrands JG, Eckenwalder JE and Richardson J (Ottawa, ON: NRC Research Press), 2001.
[63] Gordon JC. Poplars: trees of the people, trees of the future. Forestry Chron 2001 77, 217–219.
[64] Rishi AS, Nelson ND, Goyal A. Genetic modification for improvement of *Populus*. Physiol Mol Biol Plants 2001, 7, 7–21.
[65] Tuskan GA. Short-rotation woody crop supply systems in the United States: what do we know and what do we need to know? Biomass and Bioenergy 1998, 14 (4), 307–315.
[66] Wullschleger SD, Jansson S, Taylor G. Genomics and forest biology: *Populus* emerges as the perennial favorite. Plant Cell 2002, 14, 2651–2655.
[67] Sterky F, Regan S, Karlsson J, Hertzberg M, Rohde A, Holmberg A et al. Gene discovery in the wood-forming tissues of poplar: analysis of 5692 expressed sequence tags. Proc Natl Acad Sci USA 1998, 95, 13330–13335.
[68] Confalonieri M, Balestrazzi A, Bisoffi S, Carbonera D. *In vitro* culture and genetic engineering of *Populus* spp.: synergy for forest tree improvement. Plant Cell Tiss Org Cult 2003 72, 109–138.
[69] Thakur AK, Saraswat A, Srivastava DK. *In vitro* plant regeneration through direct organogenesis in *Populus deltoides* clone G48 from petiole explants J Plant Biochem Biotechnol 2012, 21(1), 23–29.
[70] Thorpe TA, Patel KR. Clonal propagation: adventitious buds. In Cell Culture and Somatic Cell Genetics of Plants. Ed. Vasil IK Vol. 1, Academic Press, Orlando 1984, 49–60.
[71] Brown DCW, Thorpe TA. Plant regeneration by organogenesis. In Cell Culture and Somatic Cell Genetics of Plants. Ed. Vasil IK Vol. 3, Academic Press, Orlando 1986, 49–65.
[72] Murashige T. Plant Propagation through Tissue Cultures. Annu Rev Plant Physiol 1994, 25, 135–166.
[73] Cavusoglu A, Altas ZI, Bajrovic K, Gozukirmizi N, Zehir A. Direct and Indirect Plant Regeneration from Various Explants of Eastern Cottonwood Clones (*Populus deltoides* Bartram ex Marsh.) with Tissue Culture. African J Biotechnol 2011, 10 (16), 3216–3221.

[74] Noel N, Leple JC, Pilate G. Optimization of *in vitro* micropropagation and regeneration for *Populus x interamericana* and *Populus x euramericana* hybrids (*P. deltoides, P. trichocarpa* and *P. nigra*). Plant Cell Rep 2002, 20, 1150–1155.
[75] Chaturvedi HC, Sharma AK, Agha BQ, Jain M, Sharma M. Production of cloned trees of *Populus deltoides* through *in vitro* regeneration of shoots from leaf, stem and root explants and their field cultivation. Indian J Biotechnol 2004, 2, 203–208.
[76] Kang B, Osburn L, Kopsell D, Tuskan GA, Cheng ZM. Micropropagation of *Populus trichocarpa* 'Nisqually-1': the genotype deriving the *Populus* reference genome. Plant Cell Tiss Organ Cult 2009, 99, 251–257.
[77] Maheshwari P, Kovalchuk I. *Agrobacterium*-mediated stable genetic transformation of *Populus angustifolia* and *Populus balsamifera*. Front Plant Sci 2016, 7, 296.
[78] Song J, Lu S, Chen ZZ, Lourenco R, Chiang VL. Genetic Transformation of *Populus trichocarpa* Genotype Nisqually-1: A Functional Genomic Tool for Woody Plants. Plant Cell Physiol 2006, 47(11), 1582–1589.
[79] Aggarwal G, Gaur A, Srivastava DK. Establishment of high frequency shoot regeneration system in Himalayan poplar (*Populus ciliata* Wall. ex Royle) from petiole explants using Thidiazuron cytokinin as plant growth regulator. J For Res 2015, 55(2), 179–188.
[80] Jiang C, Liu Z, Zheng Q. Direct regeneration of plants derived from *in vitro* cultured shoot tips and leaves of poplar (*Populus × euramericana* 'Neva'). J Life Sci 2015, 9, 366–372.
[81] Lloyd G, McCown B. Woody plant medium (WPM) – a mineral nutrient formulation for microculture of woody plant species. HortScience 1981, 16, 453–453.
[82] Driver JA, Kuniyuki AH. *In vitro* propagation of 'Paradox' walnut rootstock. HortScience 1984, 19, 507–509.
[83] Arun Kumar S, Muthuselvam M. Analysis of Phytochemical Constituents and Antimicrobial Activities of *Aloe vera* L. against Clinical Pathogens. World J Agric Sci 2009, 5, 572–576.
[84] Bhutani KK, Gohil VM. Natural products drug discovery research in India: status and appraisal. Indian J Exp Biol. 2010, 48(3), 199–207.
[85] George EF, Sherrington PD. Plant Propagation by Tissue Culture – Handbook and Directory of Commercial Laboratories. Exegetics Ltd., Edington. 1984, http://dx.doi.org/10.1002/jobm.3620250714.
[86] Van der Heijden R, Jacobs DI, Snoeijer W, Hallard D, Verpoorte R. The *Catharanthus* alkaloids: Pharmacognosy and biotechnology. Curr Med Chem 2004; 11:607–628.
[87] Jaleel CA, Gopi R, Alagulakshmanan GM, Panneerselvam R. Triadimefon induced changes in the antioxidant metabolism and ajmalicine production in *Catharanthus roseus* (L.) G. Don. Plant Sci 2006, 171, 271–276.
[88] Gayatri L, Chakravarthy R. Micropropagation in *Catharanthus roseus*. International Journal of Innovative Technology and Exploring Engineering 2013, 2, 2278–3075.
[89] Wakhlu AK and Bajwa PS. Regeneration of uniform plants from somatic embryos of *Papaver somniferum* (opium poppy). Phytomorphology, 1986, 36, 101–105.
[91] Hirata K, Yamanaka A, Kurano N, Miyamoto K, Miura Y. Production of indole alkaloids in multiple shoot culture of *Catharanthus roseus* (L.). G. Don. Agric Biol Chem 1987, 51 (5), 1311–1317.
[92] Junaid A, Mujib A, Bhat MA, Sharma MP. Somatic embryo proliferation, maturation and germination in *Catharanthus roseus*. Plant Cell Tiss Org Cult 2006, 84, 325–332.
[93] Swanberg A, Dai W. Plant Regeneration of Periwinkle (*Catharanthus roseus*) via Organogenesis. HortScience 2008, 43 (3), 832–836.
[94] Verma P, Mathur AK. *Agrobacterium tumefaciens*-mediated transgenic plant production via direct shoot bud organogenesis from pre-plasmolyzed leaf explants of *Catharanthus roseus*. Biotechnol Lett 2011, 33(5), 1053–60.

[95] Kumar A, Prakash K, Sinha RK, Kumar N. *In vitro* plant propagation of *Catharanthus roseus* and assessment of genetic fidelity of micropropagated plants by RAPD Marker Assay. Appl Biochem Biotechnol 2013,169, 894–900.
[96] Yuan YJ, Tsung, Hu T, MinYang Y. Effects of auxins and cytokinins on formation of *Catharanthus roseus* G. Don multiple shoots Plant Cell Tiss Org Cult 1994, 37, 193–196.
[97] Singh R, Kharb P, Rani K. Rapid micropropagation and callus induction of *Catharanthus roseus in vitro* using different explants. World J Agr Sci 2011, 7, 699–704.
[98] Leuang AY. *Aloe Vera* in cosmetics. Drug Cosmetics Ind 1977, 120–134.
[99] Gui YL, Xu TY, Gu SR, Lin SQ, Zhong Z, Sun GD, Zhang Q. Studies on stem tissue culture and organogenesis of *Aloe vera*. Acta Bot Sin 1990, 32, 606–610.
[100] Aggarwal D, Barna KS. Tissue culture propagation of elite plant of *Aloe vera* Linn. J Plant Biochem Biotech 2004, 13, 77–79.
[101] Chaudhary AK, Ray AK, Jha S, Mishra IN. Callus formation, shoot initiation and *in vitro* culture of *Aloe vera*. Biotechnol Bioinforma Bioeng 2011, 1(4), 551–553.
[102] Jayakrishna C, Karthik C, Barathi S, Kamalanathan D, Indra Arulselvi P. *In vitro* propagation of *Aloe barbadensis* Miller, a miracle herb. Res Plant Biol 2011, 1(5), 22–26.
[103] Reynolds T, Dweck AC. *Aloe vera* leaf gel: A review update. J Ethnopharmacol 1999, 68, 3–37.
[104] Kathi JK, Chiou V. Aloe vera (*Aloe vera*). The Longwood Herbal Task Force and The Center for Holistic Pediatric Education and Research 1999, 2, 29.
[105] Rout GR, Mahato A, Senapati SK. *In vitro* clonal propagation of *Nyctanthes arbortristis*. Biol Plant 2008, 52, 521–524.
[106] Bantawa P, Saha-Roy O, Kumar S, Ghosh R, Mondal KT. *In vitro* regeneration of an endangered medicinal plant *Picrorhiz scrophulariiflora*. Biol Plant 2011, 55, 169–172.
[107] Swarna J, Ravindhran R. *In vitro* propagation and assessment of genetic integrity of *Talinum triangulare* (Jacq.) Wild: a valuable medicinal herb. Acta Physiol Plant. 2012, 34, 1987–1996.
[108] Natali L, Sanchez IC, Cavallini A. *In vitro* culture of *Aloe barbadensis* Mill. micropropagation from vegetative meristems. Plant Cell Tiss Org Cult 1990, 20, 71–74.
[109] Meyer HJ, Staden J. Rapid *in vitro* propagation of *Aloe barhadensis* Mill. Plant Cell Tiss Org Cult 1991, 76(1), 83–86.
[110] Baksha R, Jahan MAA, Khatun R, Munshi JL. Micropropagation of *Aloe barbadensis* Mill. through *in vitro* culture of shoot tip explants. Plant Tiss Cult Biotechnol 2005, 15(2), 121–126.
[111] Hosseini R, Parsa M. Micropropagation of *Aloe vera* L. grown in south Iran. Pak J Biol Sci, 2007, 10(7), 1134–7.
[112] Hashemabadi D, Kaviani B. *In vitro* proliferation of an important medicinal plant *Aloe*- A method for rapid production. Aust J Crop Sci 2010, 4(4), 216–220.
[113] Murashige T and Skoog F (1962) A revised medium for rapid growth and bioassays with tobacco tissue culture. Plant Physiol 15: 473–497.
[114] Molsaghi M, Moieni A, Kahrizi D. Efficient protocol for rapid *Aloe vera* micropropagation. Pharm. Biol 2014, 52, 735–739.
[115] Sahoo S, Rout GR. Plant regeneration from leaf explants of *Aloe barbadensis* Mill. and genetic fidelity assessment through DNA markers. Physiol Mol Biol Plants 2014, 20, 235–240.
[116] Abrie AL, Staden JV. Micropropagation of the endangered *Aloe polyphylla*. Plant Growth Regul
[117] Chandler SF. Practical lessons in the commercialization of genetically modified plants – long vase life carnation. Acta Hort 2007, 764, 71–81.
[118] Akin Idowu PE, Ibitoye DO, Ademoyegun OT. Tissue culture as a plant production technique for horticultural crops. Afr J Biotech 2009, 8 (16), 3782–3788.

[119] Anis M, Ahmad N. Plant Tissue Culture: A Journey from Research to Commercialization. Plant Tissue Culture: Propagation, Conservation and Crop Improvement. doi 10.1007/978-981-10-1917-3_1, 2016, 3–13.
[120] Khan MI. Studies on *in vitro* morphogenesis and propagability of two species of willow (*Salix alba* L. and *Salix tetrasperma* Roxb.). PhD thesis submitted to the Aligarh MuslimUniversity, Aligarh- 202 002 (UP), India, 2014.
[121] Karppinin K, Gyorgy, Kauppinon M, Tolonen A, Jalonen J, Neubauer P, Hohtola A, Haggman, H. *In vitro* propagation of *Hypericum perforatum* L. and accumulation of hypericins, pseudohypericins and phloroglucinols. Prop Ornam Plants 2006, 6, 170–179.
[122] Tawfik AA, Read PE, Cuppett SL. Effect of some nutritional factors on monoterpene synthesis in *Rosemarinus officinalis* cultured in vitro. Acta Hort 1992, 319, 189–193.
[123] Gerth A, Schmidt D, Wilken, D. The production of plant secondary metabolites using bioreactors. Acta Hort 2007 764:95–103.
[124] Wu CH, Dewir YH, Hahn EJ, Paek KY. Optimization of the culture conditions for the biomass and phenolics production from adventitious roots of *Echinacea* angustifolia. Acta Hort 2007, 764, 187–193.

S.Vimala Devi, P.R. Kole, R. Gowthami, N. Sehrawat

# 8 Genetically modified plants: Developments and industrial aspects

**Abstract:** In the present era of genomics, the genetic modification is no longer confined to species but has rather extended across genera. The research and development in the field of biotechnology has encompassed all spheres of biological sciences, despite all legal formalities and ethical issues associated with its commercial application. Industrial biotechnology is one such field where the biotechnological process is already being used and has been applied for commercial exploitation as the genetic modification does not directly apply to the food products. The potential industries for the development and application of genetic modifications are fiber, paper and pulp, timber, biofuels, pharmaceuticals products, vaccines and other industrial oils/fatty acids. This chapter reviews some of the major biotechnological interventions in these industrial sectors.

**Keywords:** biofuels, fiber, genetic modification, industrial oils and fatty acids, paper and pulp, pharmaceuticals, timber, vaccines

## 8.1 Introduction

Genetic modification (GM) today has tremendously advanced making better the life of mankind [1]. Developments in the field of genomics and the recombinant DNA (rDNA) technology or genetic engineering have enabled the transfer of genes not only across the species but also across the genera [2–4]. It has emerged as a powerful tool with great potential for its application in the improvement of food quality as well as in the industrial supply. Its application in food crops varies with countries, as its potential hazards are considered to be individually related to the environment, biological diversity and human health. Although there is no evidence of the unique hazards with the novel combination of genes in GM crops, there is a concern particularly with the food crops and it raises suspicion among consumers and food experts [5, 6]. Many countries, including India, have implemented biosafety legislations and regulatory institutions regarding the research and use of genetically modified crops and the nutritive food ingredients derived from them [7].Though the debate on risks of GM crops still continues, the genetic modifications of plants are being carried out for industrial purposes in various countries such as China, Australia, Finland, Canada, Japan, Germany, Portugal, New Zealand, Sweden, Spain, the United States and the United Kingdom. Some of the genetically engineered plants are in fact being commercially used in the industry [8].

https://doi.org/10.1515/9783110563337-008

Industrial crops are those that are used as feedstock in the production of commercial commodities. They are not directly used for human consumption, but they contribute to the farm income in the agriculture sector [9]. They often reduce the cost of production by providing the substitute for typical imports from other countries. Genetic engineering has significant potential in many industrial sectors, namely, pharmaceuticals, biofuel, plastics, oil and lubricants, soil remediation, phytoremediation, fiber, pulp and paper, and so on [10, 11]. Genetic modification of a crop either improves its yield and quality and thus making it worth as a feedstock in the industry like timber or biofuel or it reduces the production cost of the crop by modification. As the genetically modified crops have great benefits for the commercial market, the private industry is moving toward zero waste as one of the corporate goals for environment. In this series, the most recent invasion of GM technology is engineering plant species for bioplastics that are biodegradable [12]. Some crops having industrial importance and benefits are soybean, cotton, rape seed, oil palm, safflower, peas, potato, flax, fiber hemp, agave, cassava, kenaf, guar, jojoba and so on. Some transgenic plants that can produce specific compounds for industrial applications are already commercialized by various research institutes, universities and biotechnology companies and many others are at different stages of development [13–15].

## 8.2 Developments in genetic engineering

Plants are the most significant biological resources for producing an enormous variety of chemical compounds. Genetic engineering introduces characteristic features to cultivation and postharvest to improve quality, productivity and value of important agronomic crops [11]. It also helps in enhancing nutritional value of GM crops. Various GM crops have been developed for the production of industrial products (plastics, biofuels) and recombinant medicines (vaccines, monoclonal antibodies) to treat genetic disorders [10, 12]. One of the oldest and most versatile tools in plant biotechnology is plant tissue culture, which has gained importance in recent years because of transgenic development. A vital requirement for transgenic plants is the formation of efficient somatic embryos as these originate from single cells. Somatic embryogenesis was initially induced in carrot cell suspension cultures [16, 17]. Now it has been extended to many crops, including the industrially important crops like poplar and Jatropha, and is efficiently deployed for developing transgenic [13, 18]. Genetic transformation began in early 2000 and since then many transgenic plants have been exploited in the field at commercial level. This has gained momentum because of the new range of molecular markers and the new genotype sequencing facilities. The genomics, proteomics and the advanced phenotyping facilities generate huge and precise information on the genes responsible for the critical traits, which can be harnessed in the industrial biotechnology by either over expression or down-regulating them. This has been incredibly complemented by the RNAi technology in application of gene regulation [19]. This has opened new prospects for development of transgenic for over expression or silencing of a gene as per the requirement using small interfering RNA (siRNA)or artificial micro-RNA (amiRNA)-based hairpin constructs in the industrial sector [20, 21].

## 8.3 Potential industrial applications of genetic manipulation

In the present-day agriculture, green technology has become the slogan for every commercial venture. Hence, the application of genetic engineering in plants to produce plant-based bio-products at industrial scale has gained momentum and many are already available on the market [9]. In addition to direct consumption of crops by humans, the fiber industry, paper and pulp, biofuels, pharmaceuticals, paint industry and bioplastics (Figure 8.1) are some of the areas where the genetic manipulation of plants can tremendously increase their commercial value and efficiency [20, 22, 23].

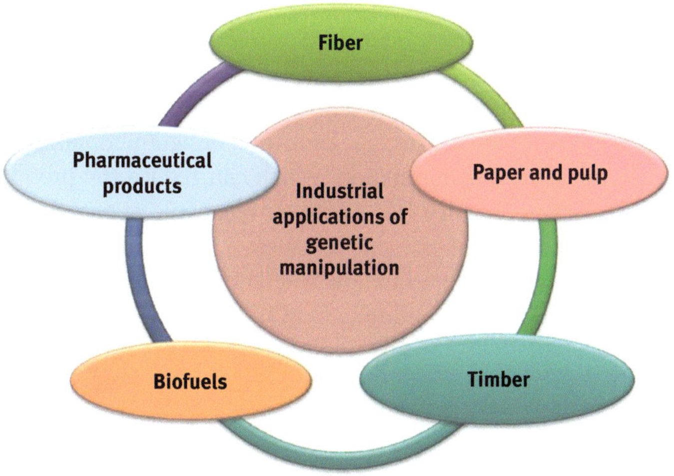


Figure 8.1: Industrial applications of genetic manipulation in various fields.

### 8.3.1 Fiber

Cotton is the most important cash crop in global agricultural industry. Bt cotton is one of the first transgenic crops to have been approved for commercial use and is considered a global revolution [21, 24]. The manipulation of the fiber crop is required not only to resist pest and diseases for high yield but also for fiber quality. However, in recent years, there has been growing demand for natural fibers and hence various approaches have been employed to improve the fibers not only in cotton but also in jute, flax and kenaff (Figure 8.2). Besides cotton, other fiber crops are grown in marginal lands that have low productivity. This may provide more space for food crops.

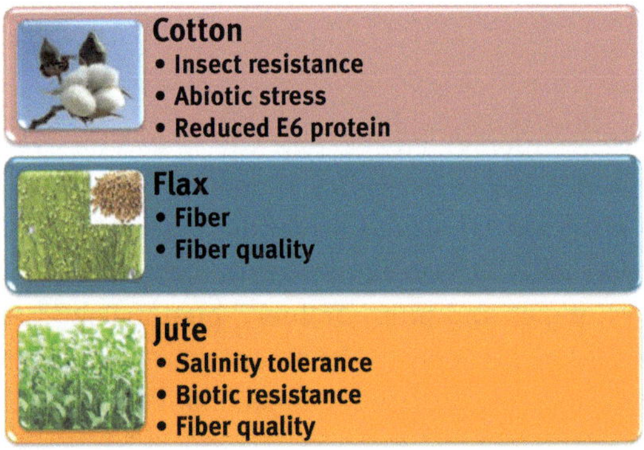


Figure 8.2: Some example of transgenic fiber crops with their modified features.

Hence, there is a great challenge to develop fiber crops with enhanced tolerance toward environmental stress [25].

Owing to the presence of biodegradable lignins, the fibers from these plants are obtained through a difficult process called retting. Reduction of lignin in plants defines a promising sphere for genetic manipulation, which can reduce the production cost and improve economic efficiency [15]. It got far-reachable application in textile industry, which was hitherto not possible because of low elasticity due to lignin. The demand for natural fibers from flax and jute is in over 30 countries of the world [26]. Some of the significant achievements in this area across the world are given in Table 8.1.

## 8.3.2 Paper and pulp

The major feedstock for the paper and pulp industry is lignin, an intricate aromatic hetero-polymer. Lignin is a key constituent of the secondary cell walls of the vascular plants [32–34]. Tree species are important resources of supply because of its huge biomass. Many billion-dollar industries thrive on the paper and pulp industry, but they also involve huge operational cost for chemical delignification in the production of high-quality paper. Many alternative genetic manipulation approaches have been made to either reduce the lignin content or alter the lignin composition (theS and Gunits, i.e., amounts of hydroxyl and methoxy groups), which help increase the pulp

Table 8.1: Some transgenic examples of the industrially important fiber plants.

| Crops | Traits | Gene | Result | Reference |
|---|---|---|---|---|
| Cotton | Insect resistance | Cry protein gene from *Bacillus thuringiensis* | Resistance against bollworms | [24] |
| | Insect resistance | Cry1Ac/Cry1Ab chimeric protein | Resistance against lepidopteran insects | [27] |
| | Insect resistance | Transgenic cotton pyramids that combine protection from Bt and RNAi). Suppression of JH acid methyltransferase (JHAMT) and JH-binding protein (JHBP) | Resistance against lepidopteran insects | [21] |
| | Abiotic stress | Cotton fiber-specific promoter (CFSP)-reporter constructs from six cotton fiber-specific genes (*Expansin, E6, Rac13, cela1, LTP*, and *Fb late*) | Fiber development under abiotic stress condition | [20] |
| | Reduced E6 protein | Genes corresponding to cotton fiber mRNA, E6 | Reduction in E6 protein for structural integrity of cotton fibers | [28] |
| | Fiber quality | Ghexpa8 fiber expansion gene | Improve fiber length and micronaire value | [25] |
| Flax | Fiber flax | Protocol of bio ballistic transformation of fiber flax, rhizogenesis induction using GFP-TUA6 construct | Protocol for transformation for fiber flax | [23] |
| | Fiber | A multigene construct containing phb A, phb B and phb C genes under the control of 14-3-3 promoter (isoform 16R) was inserted into the flax genome | PHB accumulation improved mechanical properties of flax fibers | [29] |

Table 8.1 (continued)

| Crops | Traits | Gene | Result | Reference |
|---|---|---|---|---|
| Jute | Salinity tolerance | Kate gene from *Escherichia coli* | Salt tolerance in jute for fiber | [14] |
| | Fiber quality | Down-regulation of monolignoid biosynthetic genes: cinnamate 4-hydroxylase (*C4H*) and caffeic acid *O*-methyltransferase (*COMT*) | Reducing lignin content in jute fiber | [15] |
| | Fiber quality | chs, chi, dfr gene | Increased antioxidants content | [30] |
| | Biotic resistance | B-1,3-glucanase gene | Improved resistance to *Fusarium oxysporum* and *Fusarium culmorum* | [31] |

recovery [35, 36]. This can be achieved through transgenics by increasing the S lignin content through upregulation of F5H and F5H-specific OMT, enriching lignin with cinnamaldehyde, downregulation of CAD leading to an increase in the levels of C6–C1 compounds, which helps in easy extract of lignin and others [37]. Lignin genetic engineering has been achieved in leucaena, poplar, eucalyptus and many other tree species (Table 8.2). Fast growing biotic and abiotic tolerant tree species is another aspect where genetic manipulation in tree species can directly impact this industry. Valuable traits have been introduced by genetic engineering in several crops such as insect resistance in poplar, white spruce, larch, walnut trees and eucalyptus and salinity tolerance in GM eucalyptus to grow in salt-affected areas. European black poplar is also genetically modified with Bt gene, a novel insecticide derived from the arrowhead lily, arrowhead proteinase inhibitor (API) and fast growing eucalyptus. Apart from these, the most common species are aspens, cotton woods, pines and some conifers that are being altered genetically for their use in industrial application [38].

As the pressure on the forest resources is high, use of nonwood fibrous jute, maize switchgrass, and so on as raw materials in the paper industry has been made possible with the genetic manipulation using gene regulation (Table 8.2) [39]. These nonwood resources can be produced round the year, hence ensuring round-the-year industrial supply and as proenvironment, more $CO_2$ can be absorbed by these preserved forests [14, 40].

Table 8.2: Genetic modification in some plant species for paper and pulp content.

| Crop | Trait | Gene | Result | Reference |
|---|---|---|---|---|
| Poplar | Lignin content | *C3H* RNAi | Decreased (56%) lignin content; generation of *p*-hydroxyphenyl units at the expense of G units while the proportion of S moieties remained constant | [19] |
| Poplar | Lignin composition | *OMT* antisense | Unchanged lignin content; decreased S; new 5-OH G; pale rose xylem | [41] |
| Aspen | Lignin composition | *OMT* cosuppression | Unchanged lignin content; decreased S; red brown coloration | [42] |
| Poplar | Lignin content and composition | (*OMT* sense) | Decreased (17%) lignin content; near elimination of S; increased cellulose; new 5-OH G; brownish coloration | [43] |
| *Leucaena leucocephala* | Lignin content and composition | *OMT* antisense | Decreased (28%) lignin content; decreased S; increased cellulose (9%); increased methanol-soluble phenolics; increased alkali-soluble phenolics; increased pulp extractability | [32] |
| Switchgrass | Lignin content and composition | *OMT* antisense | Decrease in lignin content; reduced recalcitrance; decreased S:G; plants; improved forage quality; increased ethanol yield (up to 38% using conventional biomass fermentation processes) | [18] |
| Poplar | Lignin content and composition | *F5H/CALD5H* overexpression | Decreased (25%) lignin content; increased S | [44] |
| Aspen | Lignin content and composition | *F5H/CALD5H* sense | Unchanged lignin content; increased S:G (threefold) | [45] |
| Poplar | Lignin content and composition | *F5H* overexpression | No significant modification in total lignin content; alteration of S lignin; significant increases in chemical (kraft) pulping efficiency from greenhouse-grown trees (decreases of 23 kappa units and increases of > 20 ISO brightness units as compared to wild-type wood); no observed phenotypic differences | [46] |

Table 8.2 (continued)

| Crop | Trait | Gene | Result | Reference |
|---|---|---|---|---|
| *Pinus radiata* | Lignin content and composition | *HCT* RNAi | Decreased (42%) lignin content; predominant deposition of 4-hydroxyphenyl units (31% of wild-type); increased resinols; reduced dibenzodioxocins; presence of glycerol end groups | [47] |
| Aspen | Lignin content | *4CL* antisense | Decreased (45%) lignin content; increased cellulose content | [48] |
| Chinese white poplar | Lignin content | *4CL* antisense | Decreased (42%) lignin content; unchanged holocellulose content; red brown coloration | [49] |
| *Populus tremula* × *Populus alba* | Lignin content | *4CL* antisense | Decreased (50%) lignin content; reddish-brown discolored wood | [36] |
| Aspen | Lignin content and composition | *4CL* × *CALD5H* antisense *4CL* and sense *CALD5H*; multigene cotransformation | Decreased (52%) lignin content; increased S:G (64%); increased cellulose (30%); accelerated cell maturation in stem secondary xylem | [45] |
| Poplar | Lignin content and composition | *CAD* antisense | Unchanged lignin content; increased aldehyde | [50] |
| Flax | Lignin content and composition | *CAD* RNAi | Decreased lignin content; changed lignin composition; reduction in the pectin and hemicellulose contents; improved mechanical properties; increased Young's modulus (75% higher); reduced (twofold) resistance of the transgenic lines to *Fusarium oxysporum* | [33] |
| Maize | Lignin content and composition | *CAD* RNAi | Unchanged lignin content, slightly decreased S:G ratio and increased cellulose and arabinoxylan contents in stems; reduced lignin content and cell wall polysaccharides in midrib; midribs and stems more degradable than wild-type plants | [51] |

### 8.3.3 Timber

Timber industry entirely depends on the forest tree species as the raw material source. Timber density – an indicator of wood quality – is amenable to improvement using gene manipulation [38]. Increase in the stiffness of core wood (*Pinus radiate*), reduced grain spiraling and desirable trunk shape are the traits that can be modified to improve the value of timber [47]. This work has been initiated with the genetic mechanism that controls timber quality like cellulose synthesis, microfibril angle, fiber cell length, cell wall thickness and early wood–latewood transitions. In addition, transgenic for resistance to pest, diseases and herbicide tolerance toward abiotic stress(freeze tolerance in eucalyptus) are important traits on which work is in progress in several research institutes for value addition to timber species[52,53].

### 8.3.4 Biofuels

As an alternative to the depleting fossil fuels, the first-generation biofuels that were made using agricultural feedstocks play a significant role in establishing the basic infrastructure for biofuel production processes and in producing a part of transport fuels. However, in several countries the competition with food crops for land use and the food security mission has eventually led to the second-generation (based on the non-food crops) and the third-generation (based on microalgae) biofuels [54]. Sequential and spatial expression of fatty acid and lipid biosynthetic genes are related with the storage lipids accumulation in the seeds of oil plants [55]. Genetic alterations of plants by transgenic technology help in developing sustainable bio-feedstocks for biofuels production. Active oilseed feedstocks can be genetically modified to yield higher oil content and optimal fatty acid composition. This can be an effective strategy to improve the yield and fuel properties of biodiesel [56]. Furthermore, the most viable option will be the lignocellulosic biomass from crops (corn stover, wheat straw, switch grass, *Miscanthus*, canary grass, giant reed, alfalfa, sweet sorghum and *Napier* grass) and various fast-growing trees (poplars, willows and *Eucalyptus*) for the production of ethanol[57].

In the oil-bearing plant species, the genes involved in the oil biosynthesis pathway is well studied and found that diacylglycerol acyltransferase (DGAT), FAD gene and others play a key role in oil production [58]. Several studies have shown that overexpression of the rate-limiting enzyme in developing seeds can enhance their oil content up to 25% (w/w). Also, the antisense technology obstructs the ability of plants to break down the fatty acids and caused an increase in oil content of plant leaves under assured growth conditions [59].

Among the non-oil-bearing source, the accessibility to renewable cellulosic feedstock is almost indefinite around the globe. Moreover, the processing of lignocellulose into biofuel and fermentable sugars needs costly and harsh pretreatments, which further reduce the cost-effectiveness [60]. The structural genes that control the cellulose or lignin production and various transcription factors acting as regulators of lignin or cellulose synthesis can be the option for genetic modification. Earlier reports highlighted that genetic alteration of the enzymes involved in lignin biosynthesis pathway specifically reduces the necessity of pretreatment processes for fermentable sugars production. For example, downregulation of six different enzymes regulating lignin biosynthetic pathway in alfalfa – $C_3H$ (4-hydroxycinnamate3-hydroxylase), $C_4H$ (cinnamate 4-hydroxylase), CCoA-OMT (*S*-adenosyl-methioninecaffeoyl-CoA/5-hydroxyferuloyl-CoA-*O*-methyltransferase), HCT (hydroxyl cinnamoyltransferase), COMT (caffeate-*O*-methyltransferase) or F5H (ferulate 5-hydroxylase) – can either reduce or abolish the needs of chemical pretreatment for the production of fermentable sugars [54,61]. The other options can be the gene regulation of the substrate-disrupting factors or modifying the features of cellulose itself for cost-effective processing. Table 8.3. provides few examples where the modification in non-oilseed bearing crops has resulted in reduced lignin or modified cellulose, which ease the bioethanol production process industrially.

### 8.3.5 Pharmaceutical products

Recent biotechnological techniques have attracted greater attention of researchers toward producing new more effective therapeutic agents by using different botanical resources. Among the different biotechnological techniques, genetic engineering of plants is one of the most desired methods than can facilitate the production of pharmacologically active proteins like blood product substitutes, vaccines, mammalian antibodies, hormones and various therapeutic agents [74, 75]. The production of proficient biopharmaceutical compounds in plants involves appropriate selection of the host plant and suitable gene expression system such as either a food crop or a non-food crop. Commercialization of plant-based biopharmaceuticals should strictly adhere to product safety issues related to the pharmaceutical workers, general public and patients [76]. This should be given high priority along with the proper control of regulatory errors during pharmaceutical compound production from plants. Plants have immense potential to produce useful compounds for mankind. Therefore, plant resources can act as a vital production system for various important biopharmaceutical compounds [77].

Transgenic plants are most commonly used to produce pharmaceuticals usually known as plant-manufactured pharmaceuticals (PMPs).Plants are an inexpensive source of protein. Moreover, plants can be easily transformed and have significant potential to produce biopharmaceutical proteins and peptides [78]. Earlier reports recognized the plants for its huge diversity and effectiveness of chemicals (cocaine, salicylic acid, taxol, morphine, etc.) having prevailing effects on human health and

Table 8.3: Milestones for the development of transgenic plants for biofuel industry.

| Crops | Traits | Gene | Result | References |
|---|---|---|---|---|
| *Populus* | Modification of the lignin biosynthetic machinery | 4-Coumarate CoA ligase (4CL1), cinnamyl alcohol dehydrogenase (CAD), caffeic acid 3-*o*-methyltransferase (COMT), *s*-adenosyl-methionine caffeoyl-CoA (CCoAOMT), ferulate 5-hydroxylase (F5H) | Different levels of lignin reduction | [49] |
| | Lignin modification | Overexpression of the pine cytosolic gene *glutamine synthetase (GS1a)* | Reduced lignin content | [62] |
| | Lignin modification | Ectopic expression of the *Eucalyptus* egmyb1 gene | Reduced lignin content | [63] |
| | Lignin modification | Downregulation of the *4CL1, COMT* and *CAD* genes | Reduced lignin content | [39] |
| | Lignin modification | Overexpression of the *R2R3-MYB* transcription factor | Reduced lignin content | [63] |
| Switch grass | Lignin modification | Overexpresses the maize *Corngrass1 (Cg1)* microRNA | Locks the plant in the juvenile stage has reduced lignin levels, 250% more starch and higher saccharification efficiency | [64] |
| Sugarcane | Lignin modification | Downregulation of the sugarcane caffeic acid *O*-methyltransferase (COMT) gene | Reduction of the recalcitrance of biomass by reducing lignin content | [54] |
| Switch grass | Lignin modification | RNA interference of Pv4CL1 | Reduced lignin content by 80% with decreased guaiacyl unit composition and improved fermentable sugar yields | [65] |

Table 8.3 (continued)

| Crops | Traits | Gene | Result | References |
|---|---|---|---|---|
| *Populus tremuloides* | Cellulose modification | Over expression of an aspen secondary wall-associated cellulose synthase (ptdcesa8) | Easy breakdown of cellulose in fermentation process | [66] |
| Sweet sorghum | Altered lignin content | Manipulating the expression of caffeoylcoa-*O*-methyltransferase (CCOAOMT) | Lignin modification for use in ethanol production | [67] |
| Alfalfa | Lignin modification | Downregulation of 4-coumarate 3-hydroxylase (C3H) | Dramatic shift in the lignin profile and consequent altered lignin structure43, causing improved digestibility | [61] |
| Switch grass | Lignin modification | Overexpression of mir156 involved in the suppression of squamosa promoter binding protein like (SPL) genes | Increase in overall biomass accumulation coupled with the potential to increase saccharification efficiency | [68] |
| Several species | Lignin modification | Over expression of R3R3- MYB4 transcription factors | Repress lignin biosynthesis in several species | [69] |
| Switch grass | Improvement of lignocellulosic feedstocks | Overexpression of pvmyb4 | Threefold increase in hydrolysis efficiency (2.6-fold) | [60] |
| Maize | Plant oil | Fungal lactase gene (EMBL accession No. U44430) | Increased expression of oil | [70] |
| Switchgrass | Lignin biosynthetic pathway | Downregulation of the caffeic acid 3-*O*-methyltransferase EC 2.1.1.68 (COMT) gene | Cell walls of transgenic plants released more constituent sugars | [71] |
| Maize | *In-planta* expression of endocellulases | Glycoside hydrolase family 5 endocellulase, E1 (Cel5A), from *Acidothermus cellulolyticus* | Will allow the enzymes to access their substrates during cell wall construction, rendering cellulose more amenable to pretreatment and enzyme digestion | [72] |

Table 8.3 (continued)

| Crops | Traits | Gene | Result | References |
|---|---|---|---|---|
| Maize | Enzyme recovery | E1 cellulase, an *endo*-beta-1,4-glucanase from *A. cellulolyticus* | Recovery of commercial amounts of enzyme at levels greater than 16% total soluble protein (TSP) in single seed | [70] |
| Sugarcane | Self-processing | Microbial genes that produce cellulose-degrading enzymes to produce | To produce self-processing plants | [73] |

physiology [79]. In recent years, plants have been significantly used in the production of animal and human oral vaccines, particularly in the developing countries where traditional vaccines are mostly used [78, 80].

### 8.3.6 Antibodies

Plant systems have been successfully explored for the expression and development of different antibodies and their derivatives. Antibodies are large and multifaceted proteins. They have enormous ability to recognize and bind to their specific molecular targets. Normally, the plants do not produce antibodies, but several studies reported that the transgenic plants expressing antibody encoding genes can form functional antibodies [81, 82]. Monoclonal antibodies are one of the significant developments of biotechnology. These are also crucial products for both diagnostic purposes and therapeutic uses. Mice have also been used to produce traditional therapeutic monoclonal antibodies. But these proteins were voluntarily recognized as foreign by the human immune system. This rigorously restricts the use of these monoclonal antibodies for therapeutic use, especially with recurring dosing. The incidence of neutralizing antibodies that inactivate the drug often precluded further therapeutic use even in the absence of serum sickness or anaphylaxis [83]. But the recombinant technologies have allowed the replacement of murine antibodies with partially humanized or chimeric antibodies. This helped in the production of fully human antibodies. Human antibodies can be obtained from mice carrying human immunoglobulin genes. These can also be produced using yeast or other gene-expression array technologies. Moreover, the recombinant technology may be used to develop an antibody gene with higher binding affinity (affinity maturation). It has also been demonstrated that the currently available recombinant antibodies exhibit increased biological activity and reduced immunogenic response as compared with the earlier monoclonal antibodies [84]. Various genetic transformation studies successfully explored the immense potential of plants to produce various recombinant proteins, such as pharmaceuticals and industrial proteins, and other secondary metabolites under different phases of clinical trials presented in Table 8.4. At present, several crops are being genetically altered to produce hormones, enzymes, vaccines, monoclonal antibodies and anticoagulant factors for the diagnosis, treatment or prevention of diseases in animals and human [75, 78, 85].

Table 8.4: Utilization of transgenic plants for the production of various kinds of antibodies.

| Plant | Genetic transformation system | Protein | Result | Reference |
|---|---|---|---|---|
| *Immunoglobulin* | | | | |
| Tobacco | *Agrobacterium*-mediated transformation | IgG specific for human creatine kinase | IgG assembly and secretion | [86] |
| | *Agrobacterium*-mediated transformation | IgG (Guy's 13) specific for *S.mutans* surface protein (SA I/II) | Comparison of glycosylation in plant- and animal-derived IgG1 | [87] |
| | *Agrobacterium*-mediated transformation | IgG (Guy's 13) specific for *S.mutans* surface protein (SA I/II) | Synthesis of full-length IgG1 | [88, 89] |
| | *Agrobacterium*-mediated transformation | Hybrid sIgA-G specific for *S.mutans* antigen II | Synthesis of secretory immunoglobin for treatment of dental caries | [85, 90] |
| *Single-chain Fv fragments* | | | | |
| Potato | *Agrobacterium*-mediated transformation | Phytochrome binding scFv | Accumulation and storage of protein in tubers | [86] |
| Tobacco | *Agrobacterium*-mediated transformation | scFv of IgG from mouse B-cell lymphoma | Treatment of non-Hodgkin's lymphoma | [91] |
| Cereals | Particle bombardment | scFvT84.66 against carcinoembryogenic antigen | Production of tumor-associated marker antigen | [83] |

## 8.3.7 Vaccines

The vaccines produced in plants can express the entire selected proteins. Therefore, the use of precise DNA encoding only specific antigenic sequences from bacteria, parasites and pathogenic viruses has received significant consideration [92]. Many food plants such as asparagus, apple, alfalfa, banana, cabbage, barley, canola, carrots, cantaloupe, cranberry, cauliflower, cucumber, flax, eggplant, grape, lettuce, kiwi, lupins, melon, maize, pea, papaya, peanut, plum, pepper, potato, rice, raspberry, soybean, service berry, squash, sweet potato, sugar beet, strawberry, sunflower, sugarcane, tomato, wheat and walnut have been transformed to produce valuable vaccines (Table 8.5) and biopharmaceuticals (Table 8.6) [93].

## 8.3.8 Other industrial oils/fatty acids

Oil-bearing tree species produce oils or hydroxyl fatty acids. These can be used to produce super quality lubricants for jet engines and also in the manufacturing of cosmetics, shampoo, antifungal compounds, plastics, paints and coating. Castor is one of such examples that are highly required in the industrial sector. However, it contains a highly poisonous compound ricin, which is more toxic than the snake venom. Castor can be genetically engineered to produce seed oils with epoxy fatty acids as vernolic acid (12, 13-epoxy-*cis*-9-octadecenoic), which is produced in *Vernonia galamensis* [110]. Epoxy oils form tremendous coatings on plastics and steel. It also helps in the preparation of improved lubricants, paints and coatings. This kind of genetic alterations can yield plant that does not produce ricin, and thus can revitalize the interest in this crop. Genetic modifications of crop plants facilitate to "knock out" the genes responsible for production of ricin, ricinine and CB-1A [111]. Castor meals free of ricin are significant in producing rations and animal feeds. Another important oil-bearing crop is Jatropha. RNA silencing of curcin gene can inhibit production of toxin and further add economic fodder value to biofuel feedstock. Enhanced expression of the stearoyl-acyl carrier protein desaturase and JcERF genes in Jatropha can exhibit increased seed oil and improved resistance toward salt stress and frost. Guar or cluster bean (*Cyamopsis tetragonoloba*), an annual legume, is an excellent source of guar gum. This has made guar a precious commodity for export. Genetic modification of ADH (alcohol dehydrogenase) gene in this valuable crop may lead to enhanced ethanol production [112].

Calendic acid is oxidatively more unstable as compared to linolenic acid. Hence, it helps in improving the drying properties of coating applications [113]. The concentration of calendic acid is lower in soybean (20–25%) than in marigold (55%). Many other biochemical pathways are actively engineered to produce useful molecules for the chemical industry. But most of them are still at R&D stage such as

Table 8.5: Production of antigens to prepare vaccines for specific disease in transgenic plants through genetic transformation systems.

| Plant species | Genetic transformation system | Disease | Vaccine antigens | Reference |
|---|---|---|---|---|
| Black-eyed bean | Cowpea mosaic virus | Foot- and-mouth disease | Foot-and-mouth disease virus VP1structural protein | [94] |
| | Cowpea mosaic virus | Mink enteritis virus | Mink enteritis virus epitope (VP2) | [74] |
| | Cowpea mosaic virus | Rhinovirus | Human rhinovirus 14 (HRV-14)and human immunodeficiency virus type (HIV-1epitopes) | [94] |
| | Cowpea mosaic virus *Agrobacterium*-mediated transformation | HIV | HIV epitope (gp120) | [95] |
| Cowpea | Cowpea mosaic virus | HIV | HIV epitope (gp41) | [96] |
| | Cowpea mosaic virus | Mucosal vaccines not requiring adjuvants | D2 peptide of fibronectin binding protein B of *Staphylococcus aureus* | [96] |
| Potato | *Agrobacterium*-mediated transformation | Autoimmune diabetes | *Vibrio cholera* toxin B subunit–human insulin fusion | [92] |
| | *Agrobacterium*-mediated transformation | Autoimmune diabetes | Glutamic acid decarboxylase | [97] |
| | *Agrobacterium*-mediated transformation | Cholera and *E. coli* diarrhea | *E. coli* heat-labile enterotoxin LT-B | [94] |
| | *Agrobacterium*-mediated transformation | Diarrhea due to Norwalk virus | Coat protein of Norwalk virus | [98] |
| | *Agrobacterium*-mediated transformation | Cholera | *V. cholerae* toxin CtoxA and CtoxB subunits | [93] |

Table 8.5 (continued)

| Plant species | Genetic transformation system | Disease | Vaccine antigens | Reference |
|---|---|---|---|---|
| Tobacco | Tmv | Cancer | c-Myc | [99] |
| | *Agrobacterium*-mediated transformation | Cholera and *E. coli* diarrhea | *E. coli* heat-labile enterotoxin LT-B | [94] |
| | *Agrobacterium*-mediated transformation | Dental caries | *Streptococcus mutans* surface protein SpaA | [94] |
| | *Agrobacterium*-mediated transformation | Diarrhea due to Norwalk virus | Coat protein of Norwalk virus | [98] |
| | *Agrobacterium*-mediated transformation | Hepatitis B | Recombinant HBsAg | [94] |
| | Tobacco mosaic virus | Hepatitis B | Murine hepatitis epitope | [99] |
| | Cowpea mosaic virus / *Agrobacterium*-mediated transformation | HIV | HIV epitope (gp120) | [95] |
| | Tobacco mosaic virus | Influenza | Hemagglutinin | [99] |
| | Tobacco mosaic virus | Malaria | Malarial B-cell epitope | [94] |
| | *Agrobacterium*-mediated transformation | Rabies | Rabies virus glycoprotein | [100] |
| Spinach | *Agrobacterium*-mediated transformation | Rabies | Rabies virus glycoprotein | [100] |

(i) tailoring of oil composition to be used as bio-based lubricants and biofuel in *Jatropha curcas* and camelina [55,112] and (ii) altered lignin composition and content to produce more efficient biofuels and other processes, including biomaterial conversion in sorghum, poplar and sugarcane [18, 34, 114].

## 8.4 Conclusions

Genetic modifications have come a long way with diverse research applications, but the full commercial potential is yet to be envisaged. The huge industry based on raw material from plant source will achieve renovations in both the production and processing technology. The chapter has dealt with some of the modifications in different industrial arena and the genes involved in modifications, which will

Table 8.6: Preparation of biopharmaceuticals in transgenic plants.

| Plants | Proteins | Genetic transformation system | Particulars | References |
|---|---|---|---|---|
| *Anticoagulants* | | | | |
| Tobacco | Human protein C (serum protease) | *Agrobacterium*-mediated transformation | Protein C pathway | [101] |
| | Human hirudin variant 2 | *Agrobacterium*-mediated transformation | Indirect thrombin inhibitors | [91] |
| Oilseed, Ethiopian mustard | Human hirudin variant 2 | *Agrobacterium*-mediated transformation | Indirect thrombin inhibitors | [91] |
| *Recombinant hormones/proteins* | | | | |
| Tobacco | Human granulocyte–macrophage colony-stimulating factor | *Agrobacterium*-mediated transformation | Neutropenia | [102] |
| | Human erythropoietin | *Agrobacterium*-mediated transformation | Anemia | [102] |
| | Human epidermal growth factor | *Agrobacterium*-mediated transformation | Wound repair/ control of cell proliferation | [102] |
| | Human hemoglobin | *Agrobacterium*-mediated transformation | Blood substitute | [103] |
| | Human homotrimeric collagen I | *Agrobacterium*-mediated transformation | Collagen | [103] |
| | Human serum albumin | *Agrobacterium*-mediated transformation | Liver cirrhosis | [102] |
| Thale cress | Human enkephalins | *Agrobacterium*-mediated transformation | Antihyperanalgesic by opiate activity | [102] |
| Turnip | Human interferon-α | *Agrobacterium*-mediated transformation | Hepatitis C and B treatment | [102] |

Table 8.6 (continued)

| Plants | Proteins | Genetic transformation system | Particulars | References |
|---|---|---|---|---|
| Oilseed | Human enkephalins | *Agrobacterium*-mediated transformation | Antihyperanalgesic by opiate activity | [102] |
| Rice | Human interferon-α | *Agrobacterium*-mediated transformation | Hepatitis C and B treatment | [102] |
| Potato | Human serum albumin | *Agrobacterium*-mediated transformation | Liver cirrhosis | [102] |
| *Protein/peptide inhibitors* | | | | |
| Rice | Human a-1-antitrypsin | Particle bombardment | Cystic fibrosis, liver disease and hemorrhage | [104] |
| Maize | Human aprotinin | Particle bombardment | Trypsin inhibitor for transplantation surgery | [105] |
| Tobacco | Angiotensin-I-converting enzyme | *Agrobacterium*-mediated transformation | Hypertension | [106] |
| Tomato | Angiotensin-I-converting enzyme | *Agrobacterium*-mediated transformation | Hypertension | [106] |
| *Nicotianabe thamiana* | α-Trichosanthin from TMV-U1 subgenomic coat protein | *Agrobacterium*-mediated transformation | HIV therapies | [107] |
| *Recombinant enzymes* | | | | |
| Tobacco | Glucocerebrosidase | *Agrobacterium*-mediated transformation | Gaucher's disease | [108] |
| *Nutraceuticals* | | | | |
| Rice | Daffodil phytoene synthase | Particle bombardment | Provitamin A deficiency | [108] |

Table 8.6 (continued)

| Plants | Proteins | Genetic transformation system | Particulars | References |
|---|---|---|---|---|
| Potato | *Amaranthus hypochondriacus* | *Agrobacterium*-mediated transformation | Amino acid deficiency | [109] |
| Sorghum | *Brown midrib* gene that encodes caffeic acid *O*-methyltrans-ferase, alignin-producing enzyme | – | Increase in fiber digestibility | [67] |

improve the efficiency cost-wise and also create wide avenues for industrial development. However, the risks involved in the application of genetic modification is still being studied and reviewed for its practical implications.

# References

[1] Key, S., Ma, J. K. C. and Drake, P. M. W. 2008. "Genetically modified plants and human health," *Journal of the Royal Society of Medicine*, 101 (6): 290–298.

[2] James, C. 2010. Executive Summary of Global Status of Commercialized Biotech/GM crops. *ISAAA* Brief No. 42. (Metro Manila; ISAAA), 23.

[3] James, C. 2011. Global status of commercialized biotech/GM crops, ISAAA Brief 43-2011. *ISAAA*, Ithaca

[4] Liu, W., Yuan, J. S. and Stewart C. N. 2013. "Advanced genetic tools for plant biotechnology," *Nature Reviews Genetics*, 14 (11): 781–793.

[5] James, C. 2001. Global review of commercialized transgenic crops, *ISAAA* Brief No. 24. International Service for the Acquisition of Agribiotech Applications (ISAAA), New York.

[6] Huang, J., Rozelle, S. D., Pray, C. E. and Wang, Q. 2002. Plant Biotechnology in China. *Science*, 295: 674–677.

[7] Lee, M. 2009. EU Regulation of GMOs. Law and Decision Making for a New Technology. *Biotechnology Regulation Series* (Massachusetts, Edward Elgar Publishing, Inc.), 274.

[8] Maghari BM. and Ardekani AM. 2011. Genetically Modified Foods and Social Concerns. *Avicenna Journal of Medical Biotechnology*, 3(3):109–117.

[9] Kumar S. and Kumar, A. 2015. "Role of genetic engineering in agriculture," *Plant Archives*, 15: 1–6.

[10] Ma, J. K. C., Drake, P. M. W. and Christou, P. 2003. The production of recombinant pharmaceutical proteins in plants. *Nature*, 4:794–805

[11] Sticklen, M. 2005. Plant genetic engineering to improve biomass characteristics for biofuels. *Curr. Opin. Biotechnol.*, 17: 315–9

[12] Conrad U. 2005. Polymers from plants to develop biodegradable plastics. *Trends Plant Sci*, 10: 511–512.

[13] McKeon, T. A. 2003. Genetically modified crops for industrial products and processes and their effects on human health. *Trends Food Sci Tech*, 14: 229–241

[14] Islami, M. S., Azam, M. S., Sharmin, S., Sajib, A. A., Alam, M. M., Reza, M. S., Ahmed, R., Khan, H. 2013. Improved salt tolerance of jute plants expressing the katE gene from Escherichia coli. *Turk J Biol*. 37: 206–211

[15] Shafrin, F.,Ferdous, A. S., Sarkar, S. K., Ahmed, R., Al- Amin, Hossain, K., Sarker, M., Rencoret, J., Gutiérrez, A., del Rio, J. C., Sanan-Mishra, N. and Khan, H. 2017. Modification of Monolignol Biosynthetic Pathway in Jute: Different Gene, Different Consequence. *Scientific Reports*, 4: 39984.

[16] Steward, F.C., Mapes, M. O., and Mears, K., (1958) Growth and organized development of cultured cells. II. Organization in cultures grown from freely suspended cells. *Am. J. Bot.*, 45: 705–708

[17] Reinert, J. 1959: Über die Kontrolle der Morphogenese und die Induktion von AdventivembryonenanGewebekulturenausKarotten. *Planta* 53: 318–333.

[18] Fu, C.,J.R. Mielenz, X. Xiao, Y. Ge, C.Y. Hamilton, M. Rodriguez, F. Chen, M. Foston, A. Ragauskas, J. Bouton, R.A. Dixon, Z.Y. Wang. 2011. Genetic manipulation of lignin reduces recalcitrance and improves ethanol production from switchgrass. *Proceedings of the National Academy of Sciences of the United States of America*, 108, pp. 3803–3808.

[19] Coleman, H. D., Park, J. Y., Nair, R., Chapple, C., and Mansfield, S. D. 2008. RNAi-mediated suppression of p-coumaroyl-CoA 3 '-hydroxylase in hybrid poplar impacts lignin deposition and soluble secondary metabolism. *Proc. Natl. Acad. Sci.USA*105: 4501–4506.

[20] Chen Junping and John J. Burke. 2015. Developing Fibre Specific Promoter-reporter transgenic lines to study the effect of abiotic stresses on fiber development in cotton. *PLoS One.;* 10(6): e0129870.

[21] Ni M, Ma W, Wang X, Gao M, Dai Y, Wei X, Zhang L, Peng Y, Chen S, Ding L, Tian Y, Li, J, Wang H, Wang X, Xu G, Guo W, Yang Y, Wu Y, Heuberger S, Tabashnik BE, Zhang T and Zhu Z. 2017. Next-generation transgenic cotton: pyramiding RNAi and Bt counters insect resistance. *Plant Biotechnology Journal*, 15: 1204–1213.

[22] Cardi T. and C. N. Stewart Jr. 2016. "Progress of targeted genome modification approaches in higher plants," Plant Cell Reports, 35 (7): 1401–1416.

[23] Lemesh V A, Guzenko EV and Emets, AI. 2010. Development of genetically modified plants of fiber flax (Linumusitatissimum) by bioballistic transformation method. agris.fao.org/agris-

[24] Flint, H.M., Henneberry, T.J., Wilson, F.D., Holguin, E, Parks, N and Buchler, R.D.1995. The effect of transgenic cotton, *Gossypium hirsutum*, L. containing *Bacillus thuringiensis* toxin gene for the control of Pink bollworm (*Pectinophora gossypiella*) and other orthropodes. *South Western Entomologist* 20(3): 281–292.

[25] Kamran, S. Bajwa, Ahmad A. Shahid, Abdul Q. Rao, Aftab Bashir, Asia Aftab and Tayyab Husnain. 2015. Stable transformation and expression of GhEXPA8 fiberexpansin gene to improve fiber length and micronaire value in cotton. *Front Plant Sci*.6: 838.

[26] Hadar Yitzhak, 2013. Sources for Lignocellulosic Raw Materials for the Production of Ethanol. In Lignocellulose Conversion (***eds***. V. Faraco). Springer-Verlag Berlin Heidelberg.

[27] Guo S, Cui H, Xu Q and Ni W. 1995. Expressive carrier with coded insect-killing protein fusion gene, and transfer gene plant. *Chinese Patent Pub*. No. CN1134981.

[28] John Me.1996. Structural characterization of genes corresponding to cotton fiber mRNA, E6: reduced E6 protein in transgenic plants by antisense gene. *Plant Mol Biol*. 30(2): 297–306.

[29] Kwiatkowska W M, Zebrowski J, Starzycki M, Oszmiański J, Szopa J (2007a) Engineering of PHB synthesis caused improved elastic properties of flax fibres. *BiotechnolProg* 23: 269–277.

[30] Lorenc-Kukuła K., Amarowicz R., Oszmiański J., Doermann P., Starzycki M., Skała J., ŻukM., Kulma A., Szopa J., 2005. Pleiotropic Effect of Phenolic Compounds Content Increases in Transgenic Flax Plant. *J. Agric. Food Chem.*, *53* (9), pp 3685–3692
[31] Kwiatkowska W M, Lorenc-Kukula K, Starzycki M, Oszmian´ski J, Ke˛pczyn´ska E, Szopa J .2004. Expression of b-1,3-glucanase in flax causes increased resistance to fungi. *PhysiolMol Plant Pathol* 65:245–256
[32] Rastogi, S. and Dwivedi, U.N. 2006. Down-regulation of lignin biosynthesis in transgenic *Leucaena leucocephala* harboring *O*-methyltransferase gene. Biotechnology Progress, 22, *pp*. 609–616
[33] Kwiatkowska, W. M., Starzycki, M., Zebrowski, J., Oszmiański, J., Szopa, J. 2007b. Lignin deficiency in transgenic flax resulted in plants with improved mechanical properties. *JBiotechnol*, 128: 919–934
[34] Bottcher, A., Cesarino, I., Brombini dos Santos, A., Vicentini, R., Sampaio Mayer, J.L., Vanholme, R., Morreel, K., Goeminne, G., Magalhães Silva Moura, J.C., Macedo Nobile, P., Carmello-Guerreiro, S.M., dos Anjos, I.A., Creste, S., Boerjan, W., Guimarães de Andrade Landell, M.G. and Mazzafera, P. 2013. Lignification in sugarcane: biochemical characterization, gene discovery, and expression analysis in two genotypes contrasting for lignin content. *Plant Physiology* 163, 1539–1557.
[35] Chen, F. and Dixon, R.A. 2007. Lignin Modification Improves Fermentable Sugar Yields for Biofuel Production. *Nature Biotechnology*, 25, 759–761. http://dx.doi.org/10.1038/nbt1316
[36] Voelker, S.L, Lachenbruch, B, Meinzer, F.C, Kitin P, Strauss, S.H. 2010. Anti-sense down regulation of 4CL expression alters lignification, tree growth and saccharification potential of field grown poplar. *Plant Physiol.* 154:874–886.
[37] Phillips RB, Hasan Jameel and Hou Min Chang, 2013. Integration of pulp and paper technology with bioethanol production. Biotechnology for Biofuels, 6:13
[38] Hely, H., Sutela, S. and Fladung, M. 2016. Genetic Engineering Contribution to Forest Tree Breeding Efforts. In Biosafety of Forest Transgenic Trees. (eds: C. Vettori *et al.*), Springer-Verlag Berlin Heidelberg.
[39] Shen, H., C. R. Poovaiah, A. Ziebell, T. J. Tschaplinski, S. Pattathil, E. Gjersing, N. Engle, R. Katahira, Y. Pu, R. Sykes, F. Chen, A. J. Ragauskas, J. R. Mielenz, M. G. Hahn, M. Davis, C. N. Stewart, Jr., and R. A. Dixon. 2013. Enhanced characteristics of genetically modified switchgrass (Panicumvirgatum L.) for high biofuel production. *Biotechnology for Biofuels*, 6: 71–85. http://dx.doi.org/10.1186/1754-6834-6-71
[40] Kwiatkowska W M, Sko´rkowska-Telichowska K, Dyminska L, Ma˛czka M, Hanuza J, Szopa J.2009. Biochemical, mechanical, and spectroscopic analyses of genetically engineered flax fibres producing bioplastic (poly-b-hydroxybutyrate). *Biotechnol. Prog.* 25:1489–1498
[41] Doorsselaere, J. V., Baucher, M., Chognot, E., Chabbert, B., Tollier, Marie-Thérèse., Petit-Conil, M., Leplé, Jean-Charles., Pilate, G., Cornu, D., Monties, B., Montagu, M. V., Inzé, D., Boerjan, W. and Jouanin, L. 1995. A novel lignin in poplar trees with a reduced caffeic acid/ 5 hydroxyferulic acid O-methyltransferase activity. *Plant J.* 5: 855–864.
[42] Tsai, C. J., Popko, J. L., Mielke, M. R., Hu, W. J., Podila, G. K., and Chiang,V. L. 1998. Suppression of O-methyltransferase gene by homologoussense transgene in quaking aspen causes red-brown wood phenotypes. *PlantPhysiol.*117:101–112.
[43] Jouanin, L., Goujon, T., de Nadaı¨, V., Martin, M.-T., Mila, I., Vallet, C., Pollet, B., Yoshinaga, A., Chabbert, B., Petit-Conil, M., Lapierre, C., 2000. Lignification in transgenic poplars with extremely reduced caffeic acid O-methyltransferase activity. *Plant Physiology*, 123: 1363–1373
[44] Franke, R., McMichael, C. M., Meyer, K., Shirley, A. M., Cusumano, J. C., and Chapple, C. 2000. Modified lignin in tobacco and poplar plants over-expressing the Arabidopsis gene encoding ferulate 5-hydroxylase. *Plant J.* 22:223–234.

[45] Li, L., Zhou, Y. H., Cheng, X. F., Sun, J. Y., Marita, J. M., Ralph, J., and Chiang, V. L. 2003. Combinatorial modification of multiple lignin traits in trees through multigenecotransformation. *Proc. Natl. Acad. Sci. USA* 100:4939–4944.
[46] Huntley, S. K., Ellis, D., Gilbert, M., Chapple, C. and Mansfield, S. D. 2003. Improved pulping efficiency in C4H-F5H transformed poplar. *J. Agricultural Food Chemistry*, 51:6178–6183.
[47] Wagner, A., J. Ralph, T. Akiyama, H. Flint, L. Phillips, K. Torr, B. Nanayakkara, L.T. Kiri. 2007. Exploring lignification in conifers by silencing hydroxycinnamoyl-CoA: shikimate hydroxycinnamoyltransferase in *Pinus radiate. Proceedings of the National Academy of Sciences of the United States of America*, 104, pp. 11856–11861
[48] Hu, W. J. *et al.* 1999.Repression of lignin biosynthesis promotes cellulose accumulation and growth in transgenic trees. *Nature Biotechnol.*17, 808–812
[49] Caihong, J, Huayan, Z, Hongzhi, W, Zhifeng X, *et al.* 2004. Obtaining the transgenic poplars with low lignin content through down regulation of 4CL. *Chi. Sci. Bull.* 49(9):905–909.
[50] Chapple C and Carpita N.1998. Plant cell walls as targets for biotechnology. *CurrOpin Plant Biol.* 1(2):179–85.
[51] Fornalé, M. Capellades, A. Encina, K. Wang, S. Irar, C. Lapierre, K. Ruel, J.P.Joseleau, J. Berenguer, P. Puigdomènech, J. Rigau, D. Caparrós-Ruiz. 2012. Altered lignin biosynthesis improves cellulosic bioethanol production in transgenic maize plants down-regulated for cinnamyl alcohol dehydrogenase. *Molecular Plant*, 5, pp. 817–830
[52] Bishop-Hurley S.L., Zabkiewicz, R.J., Grace, L., Gardner, R.C., Wagner A. and Walter C. 2001. Conifer genetic engineering: transgenic *Pinus radiata* (D. Don) and *Picea abies* (Karst) plants are resistant to the herbicide Buster. *Plant Cell Reports*, 20:235–243. DOI 10.1007/s002990100317
[53] Henderson, A. R., and Walter, C. 2006. Genetic engineering in conifer plantation forestry. *SilvaeGenetica*, 55: 253–262.
[54] Jung, J.H, Vermerris, W., Gallo, M., Fedenko, J.R, Erickson, J.E, Altpeter, F. 2013 RNA interference suppression of lignin biosynthesis increases fermentable sugar yields for biofuel.BiotechnologyJournal. (http://onlinelibrary.wiley.com/doi/10.1111/pbi.12061/abstract;jsessionid=18402879B2BA6A72B8AB429C36CAD745.f03t02?deniedAccessCustomis
[55] Kim, H.J., Silva, J.E., Vu, H.S., Mockaitis, K., Nam, J.-W., and Cahoon, E.B. 2015. Toward production of jet fuel functionality in oilseeds: identification of FatB acyl-acyl carrier protein thioesterases and evaluation of combinatorial expression strategies in Camelina seeds. *Journal of Experimental Botany* 66, 4251–265.
[56] Gu K, Chengxin Yi, DongshengTian, Jatinder Singh Sangha, Yan Hong and Zhongchao Yin, 2012. Expression of fatty acid and lipid biosynthetic genes in developing endosperm of *Jatrophacurcas*. Biotechnology for Biofuels. 5:47
[57] Feltus FA and Vandenbrink JP, 2012. Bioenergy grass feedstock: current options and prospects for trait improvement using emerging genetic, genomic, and systems biology toolkits Biotechnology for Biofuels. 5:80
[58] Turchetto-Zolet, A. C., Christoff, A. P., Kulcheski, F. R., Loss-Morais, G., Margis, R. and Margis-Pinheiro, M. 2016. Diversity and evolution of plant diacylglycerolacyltransferase (DGATs) unveiled by phylogenetic, gene structure and expression analyses. *Genet Mol Biol.* 39(4): 524–538.
[59] Dyera J. M. and Mullenb, R. T. 2008. Engineering plant oils as high-value industrial feedstocks for biorefining: the need for underpinning cell biology research. *PhysiologiaPlantarum*, 132: 11–22.
[60] Shen, H., He, X., Poovaiah, C.R., Wuddineh, W.A., Ma, J., Mann, D.G., Wang, H., Jackson, L., Tang, Y., Stewart, C.N. Jr, Chen, F. and Dixon, R. A. 2012. Functional characterization of the switchgrass (Panicumvirgatum) R2R3-MYB transcription factor PvMYB4 fo193: 121–136. (https://www.ncbi.nlm.nih.gov/pubmed/21988539)

[61] Ralph, J., Akiyama, T., Kim, H., Lu, F., Schatz, P. F., Marita, J. M., Ralph, S. A., Reddy, M. S., Chen, F. and Dixon, R.A. 2006. Effects of coumarate 3-hydroxylase down-regulation on lignin structure. *J Biol Chem*. 281(13): 8843–853.
[62] Coleman H. D., Ca´ novas, F. M. Man, H., Kirby, E. G. and Mansfield, S. D. 2012. Enhanced expression of glutamine synthetase (GS1a) confers altered fibre and wood chemistry in field grown hybrid poplar (Populustremula X alba) (717-1B4). *Plant Biotechnology Journal*, 10: 883–889
[63] Legay, S., Sivadon, P., Blervacq, A-S., Pavy, N., Baghdady, A., Tremblay, L., Levasseur, C., Ladouce, N., Lapierre, C., Se´guin, A., Hawkins, S., Mackay, J. and Grima-Pettenati, J. 2010. EgMYB1, an R2R3 MYB transcription factor from eucalyptus negatively regulates secondary cell wall formation in Arabidopsis and poplar. *New Phytologist*, 188: 774–786
[64] Chuck G. S., Tobias, C., Sun, L., Kraemer, F., Li, C., Dibble, D., Arora, R., Bragg, J. N., Vogel, J. P., Singh, S., Simmons, B. A., Pauly, M. and Hakea, S. 2011. Overexpression of the maize Corngrass1 microRNA prevents flowering, improves digestibility, and increases starch content of switchgrass. *ProcNatlAcadSciUSA*,108:17550–17555
[65] Xu, B., Escamilla-Trevino, L. L., Sathitsuksanoh, N., Shen, Z., Shen, H., Zhang, Y. H., Dixon, R. A. and Zhao, B. 2011. Silencing of 4-coumarate: coenzymeA ligase in switchgrass leads to redyields for biofuel production. *The New phytologist*,192: 611–625. (http://onlinelibrary.wiley.com/doi/10.1111/j.1469-8137.2011.03830.x/full)
[66] Joshi C. P., Shivegowda Thammannagowda, Takeshi Fujino, Ji-Qing Gou, UtkuAvci, Candace H. Haigler, Lisa M. McDonnell, Shawn D. Mansfield, Bemnet Mengesha, Nicholas C. Carpita, Darby Harris, Seth DeBolt and Gary F. Peter, 2011. Perturbation of Wood Cellulose Synthesis Causes Pleiotropic Effects in Transgenic Aspen. *Molecular Plant*, 4: 331–345
[67] Bout S and Vermerris W. 2003. A candidate-gene approach to clone the sorghum Brown midrib gene encoding caffeic acid O-methyltransferase. *Mol Genet Genomics*. 269(2): 205–14.
[68] Sunkar R and Li YF, Jagadeeswaran G. 2012. Functions of microRNAs in plant stress responses. Trends Plant Sci. Apr;17(4):196–203.
[69] Tamagnone, L.,Aerida, A. Parr, S. Mackay, F.A. CulianezMacia, K. Roberts,C. Martin.1998.The AmMYB308 and AmMYB330 transcription factors from Antirrhinum regulate phenylpropanoid and lignin biosynthesis in transgenic tobacco. Plant Cell, 10 (1998), pp. 135–154
[70] Hood EE, Love R, Lane J, Bray J, Clough R, Pappu K, Drees C, Hood KR, Yoon S, Ahmad A, Howard JA. 2007. Subcellular targeting is a key condition for high-level accumulation of cellulase protein in transgenic maize seed. *Plant Biotechnol* J. 5(6):709–19.
[71] Tschaplinski, T. J., Standaert, R. F., Engle, N. L., Martin, M. Z., Sangha, A. K., Parks, J. M., Smith, J. C., Samuel, R., Jiang, N., Pu, Y., Ragauskas, A. J., Hamilton, C. Y., Fu, C., Wang, Z-Y., Davison, B. H., Dixon R. A. and Mielenz, J. R.2012. Down-regulation of the caffeic acid *O*-methyltransferase gene in switchgrass reveals a novel monolignolanalog. *Biotechnol. Biofuels*, 5: 71
[72] Brunecky, R., Selig, M. J., Vinzant, T. B., Himmel, M. E., Lee, D., Blaylock, M. J. and Decker, S. R. 2011. *In-planta* expression of *A. cellulolyticus* Cel5A endocellulase reduces cell wall recalcitrance in tobacco and maize. *Biotechnology for Biofuels*.4:1
[73] Harrison, M.D., Geijskes, J., Coleman, H.D., Shand, K., Kinkema, M., Palupe, A., Hassall, R., Sainz, M., Lloyd, R., Miles, S., and Dale, J.L. 2011. Accumulation of recombinant cellobiohydrolase and endoglucanase in the leaves of mature transgenic sugar cane. *Plant Biotechnology Journal* 9, 884–896.
[74] Dalsgaard, K. *et al.* 1997. Plant-derived vaccine protects target animals against a viral disease. *Nat. Biotechnol.* 15, 248–252.
[75] Saklani, A. and Kutty, S.K. 2008. Plant-Derived Compounds in Clinical Trials. Drug Discovery Today, 13: 161–171. http://dx.doi.org/10.1016/j.drudis.2007.10.010

[76] Gomord, V., Chamberlin, P., Jefferis, R. and Faye, L. 2005. Biopharmaceutical Production in Plants: Problems, Solutions, and Opportunities. *Trends in Biotechnology*, 23: 559–565. http://dx.doi.org/10.1016/j.tibtech.2005.09.003
[77] Ganz, P.R. 1996. Expression of human blood proteins in transgenic plants: the cytokine GM-CSF as a model protein. In *Transgenic plants: a production system for industrial and pharmaceutical proteins.* (eds Owen, M.R.L. & Pen, J). 281–297 (John Wiley & Sons, London, UK).
[78] Union of Concerned Scientists (2006) Position Paper: Pharmaceutical and Industrial Crops. http://www.ucsusa.org
[79] Liénard, D., Sourrouille, C., Gomord, V. and Faye, L. 2007. Pharming and Transgenic Plants. *Biotechnology Annual Review*, 13: 115–147. http://dx.doi.org/10.1016/S1387-2656(07)13006-4
[80] Gomord, V. and Faye, L. 2004. Post-translational Modification of Therapeutic Proteins in Plants. *Current Opinion in Plant Biology*, 7: 171–181. http://dx.doi.org/10.1016/j.pbi.2004.01.015
[81] Ma, J.K.C. and Hein, M.B.1996, Antibody production and engineering in plants, In *Engineering plants for commercial products and applications. (eds* Collins, G.B. & Sheperd, R.J.) 72–81.
[82] Komarnytsky, S., Borisjuk, N., Yakoby, N., Garvey, A. and Raskin, I. 2006. Cosecretion of Protease Inhibitor Stabilizes Antibodies Produced by Plant Roots. *Plant Physiology*, 141: 1185–1193. http://dx.doi.org/10.1104/pp.105.074419
[83] Stoger E, Vaquero C, Torres E, Sack M, Nicholson L, Drossard J, Williams S, Keen D, Perrin Y, Christou P, Fischer R. 2000. Cereal crops as viable production and storage systems for pharmaceutical scFv antibodies. *Plant Mol Biol*; 42:583–90.
[84] Vaquero, C., Sack, M., Schuster, F., Finnern, R., Drossard, J., Schumann, D., Reimann, A., and Fischer, R. 2002. "A carcinoembryonic antigen-specific diabody produced in tobacco," *FASEB Journal*, 16 (3): 408–410.
[85] Ma, J. K., Hiatt, A., Hein, M., Vine, N.D., Wang, F., Stabila, P., van Dolleweerd, C., Mostov, K. and Lehner, T. 1995. Generation and Assembly of Secretory Antibodies in Plants. *Science*, 268: 716–719. http://dx.doi.org/10.1126/science.7732380
[86] Conrad, U., Fiedler, U., Artsaenko, O. and Phillips, J. 1998. High-level and stable accumulation of single-chain Fv antibodies in plant storage organs. *J. Plant Physiol.*152, 708–711.
[87] Cabanes-Macheteau M, Fitchette-Lainé A, Loutelier-Bourhis C, Lange C, Vine N, Ma J, Lerouge P, Faye L. 1999. N-glycosylation of a mouse IgG expressed in transgenic tobacco plants. *Glycobiology;* 9 365–372
[88] Hiatt A, Cafferkey R, Bowdish K. 1989. The production of antibodies in transgenic plants. *Nature*; 342:76–8.
[89] Ma, J.K.C. and Hiatt, A. 1996. Expressing antibodies in plants for immunotherapy. In *Transgenic plants: a production system for industrial and pharmaceutical proteins.* B *(eds* Owen, M.R.L. & Pen, P.) 229–243 John Wiley & Sons, London, UK;
[90] Ma, J.K.C. & Vine, N.D. Plant expression systems for the production of vaccines. Curr.Topics Microbiol.Immunol. 236, 275–292 (1999).
[91] Boothe, J.G., Parmenter, D.L. and Saponja J.A.1997. Molecular farming in plants: oilseeds as vehicles for the production of pharmaceutical proteins. *Drug Develop. Res.* 42, 172–181.
[92] Lam, D.M-K., Arntzen, C.J.2000. Anti-viral vaccines expressed in plants.US 05612487 (1997); Arntzen, C.J. & Lam, D.M-K. Vaccines expressed in plants.US 5914123 (1995); Arntzen, C.J. & Lam, D.M-K. Vaccines produced and administered through edible plants. US 5484719 (1996); Arntzen, C.J., Lam, D.M-K., Mason, H.S. Vaccines expressed in plants. US 6034298.
[93] Fischer R, Hoffmann K, Schillberg A, Emans N.2000. Antibody production by molecular farming in plants. J. Biol. Regul. Hemeost. Agents;14: 83–92.

[94] Tacket, C. O. and Mason, H. S. 1999. A review of oral vaccination with transgenic vegetables. Microbes Infection; 1:777–83.
[95] Mushegian, A.R. and Shepard, R.J. 1995. Genetic elements of plant viruses as tools for genetic engineering. *Microbiol. Rev.* 59, 548–578.
[96] Brennan, F.R. *et al.* 1999. Chimeric plant virus particles administered nasally or orally induce systemic and mucosal immune responses in mice. *J. Virol.* 73, 930–938.
[97] Ma, S.-W., Zhao, D.-L., Yin, Z.-Q., Mukherjee, R., Singh, B., Qin, H.-Y., Stiller, C., and Jevnikar, A. (1997). Transgenic plants expressing autoantigens fed to mice to induce oral immune tolerance. Nat. Med. 3, 793–796
[98] Dixon, R.A. & Arntzen, C.J.1997. Transgenic plant technology is entering the era of metabolic engineering. *Trends Biotechnol.*15, 441–444.
[99] Beachy, R.N., Fitchen, J.H. and Hein, M.B. 1996. Use of plant viruses for delivery of vaccine epitopes, In *Engineering plants for commercial products and applications.* (eds Collins, G.B. & Sheperd, R.J.) 43–49 (New York Academy of Sciences, NY;)
[100] McGarvey, P. B., Hammond, J., Dienelt, M. M., Hooper, D. C., Fu, Z. F., Dietzschold, B., Koprowski, H., and Michaels, F. H. (1995). Expression of the rabies virus glycoprotein in transgenic tomatoes. *BioTechnology*, 13: 1484–1487
[101] Cramer, C.L. *et al.*1996. Bioproduction of human enzymes in transgenic tobacco. In *Engineering plants for commercial products and applications.* (eds Collins, G.B.& Sheperd, R.J.) 62–71 (New York Academy of Sciences, NY; 1996).
[102] Smith, M.D. and Glick, B.R. 2000. The production of antibodies in plants. *Biotechnol. Adv.* 18, 85–89.
[103] Halling-Sørensen B, Nors Nielsen S, Lanzky PF, Ingerslev F, HoltenLützhøft HC, Jørgensen SE.1998. Occurrence, fate and effects of pharmaceutical substances in the environment – a review. *Chemosphere*, 36, 357–394.
[104] Giddings, G., Allison, G., Brooks, D. and Carter, A. 2000. Transgenic plants as factories for biopharmaceuticals. *Nature Biotechnology*, 18: 1151–1155.
[105] Zhong, G.Y., Peterson, D., Delaney, D. E., Bailey, M., Witcher, D. R., Register, J. C., Bond, D., Li, C. P., Marshall, L., Kulisek, E., Ritland, D., Meyer, T., Hood, E. E. and Howard, J. A. 1999. Commercial production of aprotinin in transgenic maize seeds. *Mol Breeding*, 5: 345–356.
[106] Hamamoto, H., Sugiyama, Y., Nakagawa, N., Hashida, E., Matsunaga, Y., Takemoto, S., Watanabe, Y., Okada, Y. 1993. A new tobacco mosaic virus vector and its use for the systematic production of angiotensin-I-converting enzyme inhibitor in transgenic tobacco and tomato. *Biotechnology* 11, 930–932.
[107] Kumagai, M. H., Turpen, T. H., Weinzettl, N., della-Cioppa, G., Turpen, A. M., Donson, J., Hilf, M. E., Grantham, G. L., Dawson, W. O. and Chow, T. P. 1993. Rapid, high-level expression of biologically active alpha-trichosanthinin transfected plants by an RNA viral vector. *Proc. Natl. Acad. Sci.USA*90, 427–430.
[108] Burkhardt, P.K., Beyer, P., Wünn, J., Klöti, A., Armstrong, G. A., Schledz, M., von Lintig, J., Potrykus, I. 1997. Transgenic rice (*Oryza sativa)* endosperm expressing daffodil (*Narcissus pseudonarcissus)* phytoene synthase accumulates phytoene, a key intermediate of provitamin a biosynthesis. *Plant J.* 11, 1071–1078.
[109] Chakraborty, S., Chakraborty, N., and Datta, A. 2000.Increased nutritive value of transgenic potato by expressing a nonallergenic seed albumin gene from *Amaranthus hypochondriacus. Proc. Natl. Acad. Sci. USA*97, 3724–2729.
[110] Cahoon, E. B. *et al.*,2002. Transgenic production of epoxy fatty acids by expression of a cytochrome P450 enzyme from Euphorbia lagascae seed. *Plant Physiol* 128(2):615–624.
[111] Auld, D. L, Rolfe, R.D., and McKeon, T.A. 2001.Development of Castor with Reduced Toxicity. *Journal of New Seeds* 3: 61–69.

[112] Kim, M.J., Yang, S.W., Mao, H.-Z., Veena, S.P., Yin, J.-L., and Chua, N.-H. 2014. Gene silencing of Sugar-dependent 1 (JcSDP1), encoding a patatin-domain triacylglycerol lipase, enhances seed oil accumulation in *Jatropha curcas*. Biotechnology for Biofuels 7, 36

[113] Cahoon, E. B., Ripp, K. G., Hall, S. E. and Kinney, A. J. 2001. Formation of Conjugated Δ8, Δ10-Double Bonds by Δ12-Oleic Acid Desaturase-Related Enzymes. Biosynthetic Origin of Calendic Acid. *The Journal of Biological Chemistry*, 276, 2637–2643. http://dx.doi.org/10.1074/jbc.M009188200

[114] Van Acker, R., Leplé, J.-C., Aerts, D., Storme, V., Goeminne, G., Ivens, B., Légée, F., Lapierre, C., Piens, K., Van Montagu, M.C.E., Santoro, N., Foster, C.E., Ralph, J., Soetaert, W., Pilate, G. and Boerjan, W. 2014. Improved saccharification and ethanol yield from field-grown transgenic poplar deficient in cinnamoyl-CoA reductase. *Proceedings of the National Academy of Sciences*, USA 111, 845–850.

Aman Verma

# 9 Chloroplast genetic engineering: Concept and industrial applications

**Abstract:** Chloroplast engineering has emerged as an environment-friendly tool and is rather favored over nuclear engineering in some crops. Genome of plastid is highly polyploidy, and therefore it offers introduction of multiple copies of foreign gene in plant cells. Transformation of chloroplast has many inherent advantages. These advantages include higher level of foreign gene expression, multi-gene engineering in a single transformation process, absence of gene silencing and position effect variation, minimal outcrossing of transgene and precise regulation and sequestration of foreign protein. In recent years, successful chloroplast genome engineering has resulted in improved resistance to insect disease, herbicides and drought and production of biopharmaceuticals, including vaccine antigen. Furthermore, development and advancements in chloroplast transformation will help in genetic modification, genetic improvements in plants and cost-effective production of pharmaceutical components with ecofriendly approach. This chapter focuses on basic concepts of genetic engineering and its commercial prospects.

**Keywords:** chloroplast, crop improvement, genetic transformation, therapeutic proteins, transgene

## 9.1 Introduction

The human population is increasing continuously across the world and it is expected to be around 9.2 billion by the year 2050. In order to feed this increasing human burden on earth, food production should grow in parallel. Human largely depends on agriculture, and hence there is a concern for its growth. The major agriculture production systems aim at producing food for human utilization, fibers for fuel and feed for animals, as well as producing raw materials for other products, including colors, perfumes and drugs. Sustainable agriculture is the need of time to support both the existing and the future generations. Agriculture is said to be "sustainable" when it can continually support production at levels needed for payback (cash economy) or persistence (subsistence economy) [1]. The primary concern is that despite arable land being utilized to its maximum (High Level Expert Forum, FAO, October 2009) [2], the demands are still not met. Advances in the agricultural biotechnology arena, mainly plant genetic engineering, are believed to revitalize crop productivity. Plants have three main reservoirs of DNA: nucleus, mitochondria and chloroplast. Therefore, they are potential targets for genetic engineering. There are considerable researches pertaining to insert transgenes into the plastid genome

https://doi.org/10.1515/9783110563337-009

in spite of the nuclear genome. Chloroplast transformation has gained significant momentum as a new tool for transgene containment due to the maternal inheritance. This chapter describes recent advances in plastid transformation in agricultural sector. It laid emphasis on novel tools for plastid genome engineering, including expression of transgene to explore the potential of plastid transformation and its industrial relevance.

## 9.2 General features of genetic transformation

Plant biotechnologists have adopted a wide range of methodology to integrate foreign DNA into plant cells. The selection of method depends upon the objectives. Each method has its own merits and demerits. The methods have been broadly classified into two categories: (a) direct DNA transfer methods; (b) indirect DNA transfer methods. The various DNA transfer methods have been listed in Table 9.1.

Table 9.1: Various DNA transfer methods used for genetic transformation.

| S. No. | Name of the method | Merits | Demerits | References |
|---|---|---|---|---|
| 1 | *Agrobacterium*-mediated transformation | – Most extensively used method<br>– Has broad host range (mainly for dicotyledonous plants)<br>– Stable integration | – Not applicable for monocotyledonous plants | [3–6] |
| 2 | Sonication-assisted *Agrobacterium* mediated transformation (SAAT) | – Besides transformation used as transient expression assay | – Ultrasound facilitates required | [7, 8] |
| 3 | Microparticle bombardment method (biolistic gun) | – Widely adopted for plants recalcitrant to transformation with *Agrobacterium*<br>– Capable of delivering DNA into the nucleus, mitochondria and chloroplasts | – Cell damage<br>– Inability of the transformed tissue to regenerate<br>– Expensive equipment | [9–11] |
| 4 | Agrolistic transformation | – Combination of *Agrobacterium*-mediated transformation and particle bombardment method<br>– Co-transformation to deliver a gene of interest | – Expensive equipment<br>– Labor-intensive | [12, 13] |

Table 9.1 (continued)

| S. No. | Name of the method | Merits | Demerits | References |
|---|---|---|---|---|
| 5 | Electroporation | – A large electric pulse temporarily allows polar molecules like DNA to pass into the cell<br>– Generate transgenic plants by protoplast transformation<br>– Used in intact plant tissues | – Cell damage by electric pulses of wrong length/ strength<br>– Nonspecific transport during electropermeability<br>– Ionic imbalance and cell death | [14–17] |
| 6 | Microinjection | – Micromanipulator is used for introduction of DNA into the nucleus or cytoplasm | – Causes release of hydrolases and toxic compounds from the vacuole to the cytoplasm resulting in death of the protoplast | [18, 19] |
| 7 | Liposome-mediated transformation | – Microscopic spherical vesicles having DNA insert<br>– Used for protoplast transfection | – It is laborious and less efficient | [20–22] |
| 8 | Pollen-tube pathway (PTP) method | – Foreign DNA is applied to cut styles shortly after pollination<br>– DNA reaches the ovule by flowing down the pollen tube | – Low efficiency of transformation | [23–25] |
| 9 | Silicon carbide Whisker-mediated method | – Vortex silicon carbide crystals in liquid medium containing DNA and plant cells<br>– Physical piercing of the cell wall by crystal facilitating entry of DNA insert | – Low transformation efficiency<br>– Lower regeneration capacity of transformants | [26–28] |
| 10 | Nanoparticle-mediated method | – Nanoparticles are combined with chemical compounds that are difficult for cells to internalize, and these particles are then used to deliver genes into target host cells | – Mainly used in animal cell transfection | [29–31] |

## 9.2.1 Chloroplast engineering

Chloroplast engineering has emerged as a new tool and is rather favored over nuclear engineering in many cases. A comparative account of chloroplast engineering with respect to use of nuclear genome is shown in Table 9.2. The operon system (cluster of genes with a single region of promoter), which usually encodes enzymes and regulatory proteins of the same biosynthetic processes, is universal in prokaryotic organisms. The single transcript of whole multigene operon conveys information required for the formation of several proteins of the same pathway. In contrast, nuclear genes are transcribed singly and they are not aligned in operons. Moreover, genes of the chloroplast are often present in operons. Chloroplast engineering is also known as chloroplast transformation technology (CTT) and it is very popular in current scenario. Surprisingly, a surplus of 130 genes from different sources have been inserted, integrated and expressed successfully through CTT for different applications in important plants, including tobacco, rice, barley, maize, carrot, pine, potato, sunflower and cotton [32]. The eukaryotic plastid genomes database can be accessed at NCBI website. Researchers are also using CTT for developing resistance in crop plants against various biotic and abiotic stresses, namely, insects, pests, viral, fungal and bacterial pathogens, numerous types of herbicides, drought, cold and salt tolerance. The technology is also being used for developing phytoremediation of toxic metals, cytoplasmic male sterility, metabolic engineering, production of vaccine antigens and even for improved biofuels, biopharmaceutical agents and industrially important enzymes [33-37]. Moreover, the plastid genome transformation has many inherent values [38].

Table 9.2: Comparative advantages of chloroplast genome over nuclear genome.

| S. No. | Chloroplast genome | Nuclear genome |
|---|---|---|
| 1 | Reduced gene dispersal due to maternal inheritance | Gene dispersal is high in the environment due to its parental nature |
| 2 | Around 60 copies of a single circular chromosome per plastid | Two copies of each of many chromosomes |
| 3 | Approximately 50–60 chloroplasts per cell | Number of chromosomes per diploid cell is species specific |
| 4 | Each gene is separate and is transcribed individually | Many genes are in operons and are transcribed together |
| 5 | Higher expression and accumulation of foreign proteins | Lower expression and accumulation of foreign proteins |
| 6 | Efficient multiple gene expression in single transformation event | The efficiency of single transformation for multiple gene expression is very poor |

Table 9.2 (continued)

| S. No. | Chloroplast genome | Nuclear genome |
|---|---|---|
| 7 | A single promoter for expression of multi-subunit complex protein | Several promoters for each genes to drive expression of respective subunits |
| 8 | Homologous recombination avoids position effects and gene silencing | Random integration presents position effects and gene silencing |

These advantages (Figure 9.1) include higher levels of transgene expression, capacity to express polycistronic messages from a promoter, uniparental plastid gene inheritance to prevent pollen transmission of new DNA, absence of gene silencing and position effect variation, homologous recombination-mediated targeted gene renewal, specific transgenic control and seizing of foreign proteins in the cellular organelle to avoid its unpropitious interactions with the cytoplasmic domain [38].

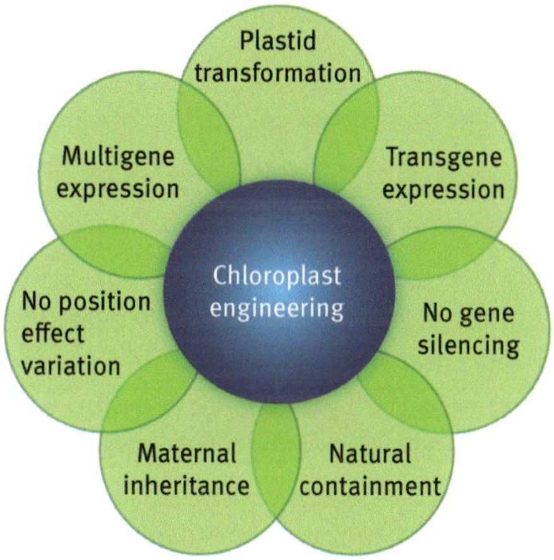


Figure 9.1: Advantages of chloroplast engineering.

## 9.2.2 Chloroplast and its genome

Plastids are the site of manufacture and storage of important chemical components used by the cell. Plastids are of the following types: (1) green-colored chloroplasts containing chlorophyll; (2) yellow-, orange- or red-colored chromoplasts containing carotenoid; (3) amyloplasts storing starch granules; (4) elaioplasts containing oil; (5) proplastids (precursor of plastid found in many plant cells); (6) etioplasts (partially matured chloroplasts that are found in dark-grown seedlings). The chloroplast is the site of manufacturing of many biomolecules like sugars, starch, complex organic compounds and many amino acids. Reduction of nitrate to ammonium and assembling chloroplast proteins are the added functional aspects of this organelle.

Plant's chloroplast and of algal cells were evolved from photosynthetic bacteria residing inside the primitive antecedents of plants [39-41]. Chloroplast possesses multiple copies of circular genome with as many as 100–250 genes. Its genomic size varies from species to species, ranging from 107 kb of *Cathaya argyrophylla* to 218 kb of *Pelargonium* and it follows maternal inheritance in angiosperm plants [37, 42]. The copy number of genome per cell depends on the plant cell age, ranging from 1000 to 10,000. For instance, a typical photosynthetic cell consists of approximately 100 chloroplasts, and each chloroplast possesses about 100 copies of DNA. These genomes happen to be in multiple subgenomic circles and small circles are originated from larger ones through crossing over. Various genes in the chloroplast genome are aligned into polycistronic transcription units, that is, cassette of two or more genes that are transcribed by RNA polymerase enzyme from a single promoter. Epigenetic effects such as gene silencing [43, 44] and position effects [45] are not present in chloroplasts because of site-specific integration of the transgenes into the spacer regions of the genome. During CTT, gene containment occurs due to the lack of transgene transmission by pollen. During pollen maturation, plastid DNA is lost and therefore not transferred to the coming generation [46].

### 9.2.3 Methods for transforming plastids

Plastid transformation was first accomplished in *Chlamydomonas reinhardtii*, a unicellular alga [47]. It was carried out for the deletion *atpB* gene of the chloroplast that rendered the green algae capable of carrying out photosynthesis and could be recovered by integrating the wild-type gene. The genomic DNA was coated on a tungsten microprojectile (~1 mm in size) and pushed into cells inoculated on an agar plate using a gunpowder charge [48]. Compressed gases were used for biolistic transformations of the host genome [49]. The DNA was thus incorporated into the chloroplast chromosome of this green alga by homologous recombination. Perhaps, tobacco was the first higher plant in which chloroplast transformation was successfully achieved [10, 50]. Later on, DNA was introduced into chloroplasts of the cultured cells of tobacco in situ or in leaf tissue and partial expression was observed. The transformed cells were further regenerated into whole plants with these modified chloroplasts [10, 51]. The basic layout for chloroplast engineering and its applications have been shown in Figure 9.2.

Major attention of biotechnologists was driven by the experiments of McBride et al. [52]. They introduced the gene using the microprojectile bombardment method to encode the *Bacillus thuringiensis* lepidopteran protoxin into the chloroplast's chromosomes of tobacco. Transgenic tobacco showed 2–3% of the soluble leaf protein expressing protoxin. Two additional approaches were used to deliver

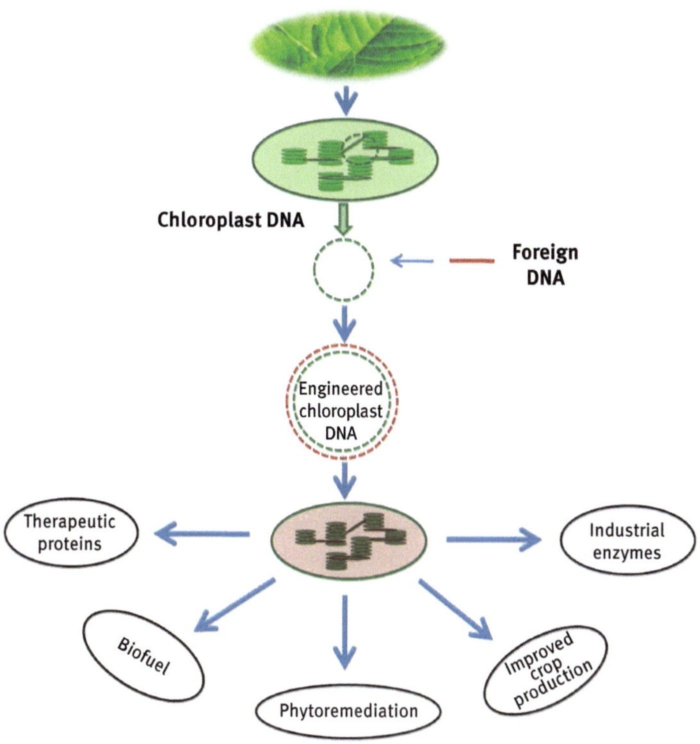


Figure 9.2: Chloroplast engineering: Basic layout and its applications.

DNA into plastids: polyethylene glycol (PEG)-mediated and direct transfer of DNA *in situ*. PEG-mediated approach was used in case of *Nicotiana plumaginifolia* [53] and *Nicotiana tabacum* [54] and it has the capacity to form more cells with genetically transformed plastids more easily than by the biolistic method. Knoblauch et al. [55] have shown another approach of direct injecting DNA into chloroplast genome in photosynthetically active leaf cells of tobacco. These findings mark the beginning of chloroplast engineering and opened new vistas for researchers to explore the potential of chloroplast. Chloroplast engineering demands some of the following gene features: (a) promoter region, (b) downstream box and N-terminal amino acid sequence for the codon usage, (c) 5′ untranslated region (5′ UTR) and (d) 3′ UTR. A brief description of the above cited regions is given below:

#### 9.2.3.1 Promoter region

The process of transcription in plastid is mediated by the controlled and combined actions of two types of RNA polymerases (RNA Pol) identifying distinct promoters:

(1) T7-like single-subunit nuclear-encoded polymerase (NEP); (2) bacterium-like α2ββ′ plastid-encoded polymerase (PEP). A characteristic feature of NEP is that it catalyzes transcription in undifferentiated plastids and in certain etiolated (non-green) tissues to produce rRNAs and mRNAs, which encode for many ribosomal proteins included in the PEP also. This however leads to the amassing of active and functional PEP. Many researchers have noticed that most of the promoters of the plastids have both PEP and NEP at transcription start sites [56]. One such example is of transcription of transgenes inserted into the plastid genome using 16S rRNA promoter (P*rrn*16) or the *psbA* promoter (P*psbA*). P*psbA* contains transcription start sites for PEP, whereas P*rrn* contains transcription start sites for both PEP and NEP [57]. In another set of experiment, a synthetic promoter system had been created by changing the P*rrn* promoter to incorporate *lac* operator sequences from the *Escherichia coli lac* operon [58]. A T7 RNA polymerase hybrid transcription system unlike NEP has been developed to synthesize bioplastic, namely, polyhydroxybutyric acid (PHB) in the plastids [59]. Many researchers have used promoterless constructs for chloroplast engineering. A read-through transcription from the original plastid ribosomal or *psbA* promoters is advantageous to achieve high-level protein expression in transformed plastids [60-62].

#### 9.2.3.2 Downstream box

The region of downstream box (DB) is of 10–15 codons, present downstream of the start site. It was first recognized in *E. coli* [66] and found to have major effects on accumulation of foreign proteins. It functions symbiotically with the Shine–Dalgarno sequences present upstream of the start site to regulate transgene protein expression and accumulation. The last decade has witnessed many studies pertaining to the influence of DB region on the alterations of transgene protein accumulation [67, 68]. Gray et al. [61] showed that the choice of DB whether native or recombinant (fused) alters the transgene expression as well as magnitude. They reported that when 14 amino acid fusions from the N-terminal of TetC DB and NPTII DB or GFP DB were fused to either an endoglucanase gene or a β-glucosidasegene from *Thermobifida fusca*, the accumulation of the transgene enzymes varied almost twice the magnitude, depending upon the type of DB sequence used. For instance, TetC DB was found to be ideal for the endoglucanase gene, while NPTII DB was found to be more expressive with the β-glucosidase gene. These findings prove that choice of DB sequence should be optimized and standardized because different outcomes may be witnessed depending upon the particular coding region that is under its influence.

#### 9.2.3.3 5′ Untranslated regions (5′ UTRs)

Since certain chloroplast genes are modulated at the posttranscriptional level [63], the integration of particular 5′ UTR into a plastid transgene may also give additional regulatory control of transgene protein expression specially designed to aim many biological processes like photosynthesis. Most regularly used 5′ UTRs are *psbA* gene, *rbcL* and the bacteriophage T7 gene 10 of the plastids. Incorporation of 5′ UTR of the bacteriophage T7gene 10 into the constructs exhibited increased expression of transgenic protein [64]. Yang et al. observed that when plants were grown in the light, the *psbA* 5′ UTR resulted in higher accumulation of β-glucuronidase protein encoded by *uidA* gene under the transcriptional control of *psbA* promoter. They also observed that when two additional bacteriophage 5′ UTRs along with *aadA* marker gene were integrated, it resulted in sufficient expression for the recovery of transgene protein [65].

#### 9.2.3.4 3′ Untranslated regions (3′ UTRs)

The 3′ UTRs are located downstream of the stop codon in the plastid genome. It consists of hairpin-loop-like structure that renders RNA maturation and posttranscriptional processing. It also checks degradation of the RNA by specific ribonucleases [69]. The 3′ UTRs are majorly derived from plastid genes, with the *rbcL, rpl32, rps16* and *psbA*, to control foreign gene expression in plastids. Like DB sequences, 3′ UTRs should be selected carefully. In CTT, the integration of foreign genes is achieved by placing it into the intergenic spacer regions keeping in mind that the original chloroplast genes remain intact. A minimum of two flanking chloroplast genes mark the site for the integration of transgene cassettes. A transgene cassette comprises a selectable marker, such as gene for streptomycin resistance and gene(s) of interest, both under the transcriptional control of specific promoters and 3′ and 5′ untranslated regions. Knowledge of chloroplast genome sequences is vital in formulating transgene cassettes as it requires sequence information of both regulatory and flanking sequences for stable integration. Chloroplast vectors are constructed by integrating transgene cassettes into suitable bacterial plasmids, which are then bombarded into plant cells by using methods of gene transfer such as biolistic gene gun. Transgene cassettes may find its way to either nuclear or mitochondrial genome, via homologous or nonhomologous recombination events. Nontargeted integration could be easily identified by evaluation of their integration site and subsequent elimination [70]. Perhaps any nontargeted integration will not result in the transgenes expression at the nuclear or mitochondrial genome as chloroplast regulatory sequences are not functional there.

Scientific community has successfully integrated many genes on chloroplasts to achieve agronomic traits of interest. For instance, transgenic tobacco has been engineered to combat multiple biotic and abiotic stresses with co-integration and expression of protease inhibitors and chitinase [71]. Plastid genome transformation has proved to be one of the leading options in achieving various economically important agronomic traits such as insect resistance, herbicide resistance and tolerance to drought, freezing/chilling and salt [72]. Herbicide tolerance (glyphosate tolerance) is the dominant trait that attracted the most for plastid genome transformation [51, 73, 74]. Many of such reports are summarized in Table 9.3. In addition to aiming for the improvement of agronomic traits, researchers are also focusing on the production of vaccines, therapeutic proteins and biomaterials because of the maternal inheritance of the chloroplast genes. Another important benefit associated with CTT is that one can avoid inserting antibiotic-resistant genes as selectable marker as certain chloroplast genes can act as the same. This will certainly improve public perception toward plant biotechnology and its products. One such gene is *betaine aldehyde dehydrogenase* (*BADH*) that can confer antibiotic-free selection of transformed plastids. Such an approach prevents arguments against genetically modified (GM) plants concerning about horizontal gene transfer from transgenic plants back to bacteria, which may develop antibiotic resistance.

Table 9.3: Applications of chloroplast genome engineering.

| Transgenes | Efficiency of expression | Engineered traits or products | Reference |
|---|---|---|---|
| Insect or pathogen tolerance | | | |
| *Pta* | 7.1–9.2% Total soluble protein (TSP) | Resistance against aphid, whitefly, Lepidopteran insects, bacterial and viral pathogens | [75] |
| *Bgl-1* | >160-fold enzyme | Resistance against aphid and whitefly | [76] |
| *RC1011, PG12* | 17–38% TSP | Resistance to *Erwinia* soft rot and tobacco mosaic virus | [77] |
| *Bt cry2Aa2* operon | 45.3% TSP | Cuboidal Bt crystals formation; 100% mortality of cotton bollworm, beet armyworm | [43] |
| *msi-99* | 21–43% TSP | Resistance to *in planta* challenge of *Verticillium dahlia, Pseudomonas syringae, Aspergillus flavus, Fusarium moniliforme* and *Colletotrichum destructivum* | [78] |
| *Bt cry9Aa2* | ~10% of TSP | Resistance to *Phthorimaea operculella* | [60] |
| *cry2Aa2* | 2–3% of TSP | Resistance to *Helicoverpa zea, Heliothis virescens* and *Spodoptera exigua* | [79] |

Table 9.3 (continued)

| Transgenes | Efficiency of expression | Engineered traits or products | Reference |
|---|---|---|---|
| Abiotic stress tolerance | | | |
| *Badh* | 93–101 /g Fresh weight (FW) | Salt tolerance: carrot plants survived up to 400 mM NaCl | [80] |
| *tps1* | >169-fold transcript | Drought tolerance: growth in 6% polyethylene glycol (PEG) and rehydration after 24 days of drought | [44] |
| *γ-TMT* | >7.7% TSP | Increased salt and heavy metal tolerance, enhanced accumulation of α-tocopherol in seeds | [81] |
| *b-bar1* | >7% TSP | Resistance to phosphinothricin (herbicide) | [82] |
| *EPSPS* | >10% TSP | Resistance to glyphosate (herbicide) | [67] |
| *panD* | >4-fold β-alanine | Tolerance to high-temperature stress | [83] |
| *Lycopene β-cyclase* | 0.28 mg/g DW | Herbicide resistance and triggers conversion of lycopene | [84] |
| *HTP, TCY, TMT* | NR | Increase in vitamin E in fruit; cold-stress tolerance | [85] |
| Other agronomic traits | | | |
| *RbcS* | >150-fold RbcS transcript | Restoration of Rubisco activity in *rbcS* mutants | [86] |
| *TC, γ –TMT* | 3 nmol/(h mg) FW | Vitamin E accumulation in lettuce and tobacco | [87] |
| *BicA* | ~0.1% TSP | $CO_2$ capture within leaf chloroplasts | [88] |
| *Bgl-1* | 44.4 units Bgl1 $g^{-1}$ FW | β-Glucosidase increased enzyme cocktail efficiently to release sugar from paper, wood and citrus peel | [76] |
| *ubiC* | 25% DW | 250-fold higher pHBA polymer accumulation than nuclear transgenic lines | [89] |
| *Cutinase* or *swollenin* | 47.7% reduction of MGDG and DGDG in cutinase and 68.5% in swollenin | Swollenin enlarged and irreversibly unwound cotton fiber; used in enzyme cocktails; cutinase showed esterase and lipase activity | [90] |
| *CV-N* | ~0.3% TSP | Increased mRNA stability and protein stability with the expression of CV-N in chloroplasts | [91] |
| *Cel6A,Cel6B* | 2–4% TSP | Hydrolyzed crystalline cellulose | [92] |

Table 9.3 (continued)

| Transgenes | Efficiency of expression | Engineered traits or products | Reference |
|---|---|---|---|
| *bgl1C, cel6B, cel9A, xeg74* | 5–40% TSP | Cell wall-degrading enzyme activity | [93] |
| *crtZ, crtW* | >0.5% DW | Astaxanthin accumulation | [94] |
| *EGPh* | 25% TSP | Chloroplast-derived β-1,4-endoglucanase (EGPh) was recovered from dry leaves and digested with carboxymethyl cellulose (CMC) substrate | [95] |
| *FVIII* | 370 mg/g FW | Feeding of the HC/C2 antigen mixture substantially suppressed T-helper cell responses and inhibitor formation against FVIII in hemophilia A mice | [96] |
| *Proinsulin* | 47% TSP in tobacco, 53% TLP in lettuce | Oral delivery of proinsulin in plant cells lowered glucose levels comparably to injectable commercial insulin | [97] |
| *IGF* | 32.7% TSP | Promoted growth of cultured HU-3 cells in a dose-dependent manner | [36] |
| *BACE* | 2.0% TSP | Immunogenic response against the BACE antigen in mice | [98] |
| *IFNα2b* | 3 mg/g FW | Protected cells against VSV CPE and HIV; increased MHC I antibody on splenocytes and the total number of natural killer cells and protected mice from a highly metastatic lung tumor | [99] |
| *IFN-γ* | 6% TSP | Protection of human lung carcinoma cells against infection by encephalomyocarditis virus | [100] |
| *CTB-2 L21* | 31.1% TSP | Immunogenic in mice following IP or oral administration | [101] |
| *LTB* | 2.5% TSP | GM1 ganglioside binding assay | [74] |
| *DPT* | 0.8% TSP | Immunogenic in orally inoculated mice with freeze-dried chloroplast-derived multi-epitope DPT protein | [102] |
| *HEV E2* | 1.09 ng/μg TSP | Immune response in mice against hepatitis E virus | [103] |
| *p24* | ~4% TSP | Induced strong CD4+ and CD8+ T-cell responses in mice | [104] |

## 9.2.4 Applications of chloroplast engineering

### 9.2.4.1 Production of vaccine antigens and biopharmaceuticals

Diabetes mellitus is a chronic disease whose prevalence is continuously increasing worldwide. The World Health Organization has estimated that around 1.5 million deaths had occurred in 2012 due to this disease, making it the eighth leading cause of death. Most of the world's population cannot afford its cure, that is, insulin, due to its high cost [37]. The production cost of insulin is high due to its requirements like fermentation systems, its expensive purification procedures from the host proteins and its costly storage and transportation. Shorter shelf-life of the finished product further adds to its cost [105, 106]. CTT can address this issue very well as it does not require such expensive production procedures and can be stored without losing the efficacy of the finished product [107, 108]. Table 9.4 has summarized many such researches aiming for engineering various vaccine antigens and biopharmaceuticals via targeting chloroplast genome of higher plants. Almaraz-Delgado et al. [109] engineered tobacco plastid for anticancer vaccine development by expressing *E7 HPV* type 16 protein. As already mentioned, CTT is preferred nowadays over nuclear transformation. One of the reasons behind this is the liberty for carrying out many posttranslation modifications easily such as phosphorylation, acetylation/deacetylation, amidation, formation

Table 9.4: Vaccine antigens, biopharmaceuticals and therapeutic proteins engineered via chloroplast genome of higher plants.

| S. No. | Traits | Gene | Expression level | Reference |
|---|---|---|---|---|
| 1 | HIV/AIDS | *gp120, gp41* | 16 µg/g FW | [116] |
| 2 | Polio virus | *CTB-VP1* | 4–5% TSP | [113] |
| 3 | Tuberculosis antigens | *CTB-SAT6CTB-Mtb72F* | 7.5% TSP | [117] |
| 4 | Dengue virus | *EDIII* | 0.8–1.6 TSP | [118] |
| 5 | Pompe disease | *CTB-GAA* | 0.1–0.2 TLP | [119] |
| 6 | Insulin liken growth factors | *IGF-1n* | 32% TSP | [118] |
| 7 | Interferon-α2b(IFN-α2b) | *IFN-a2b* | 21% TSP | [120] |
| 8 | Basic fibroblast growth factor (bFGF) | *bFGF* | 0.1% TSP | [121] |
| 9 | *Escherichia coli* phytase gene | *appA* | Not detected | [122] |
| 10 | Human glutamic acid decarboxylase | *hGAD65* | 0.25–0.3% TSP | [123] |

Table 9.4 (continued)

| S. No. | Traits | Gene | Expression level | Reference |
|---|---|---|---|---|
| 11 | Infectious burial disease virus | *IBDV-VP2* | 0.8–4% TCP | [124] |
| 12 | Mammary-associated serum amyloid | *M-SSA* | 3–5% TSP | [114] |

of disulfide bonds, proper folding and the assembly of complex multimeric proteins. Another advantageous feature is the higher level of vaccine antigens or therapeutic proteins expression in green (leaves) or nongreen (fruits/roots) chloroplasts. It also provides an antibiotic-free selection system that is best suited for oral delivery of vaccine antigens against many bacterial and viral diseases like cholera, anthrax, plague, tetanus and canine parvovirus [108, 110–112]. Lakshmi et al. [113] discovered cost-effective vaccine antigens production against tuberculosis and studied their expression in chloroplasts. While working on cancer therapeutics, Tran et al. observed increased expression of foreign proteins in chloroplasts owing to its inherent capacity, which may result in toxicity on the host plant. Temporary immersion bioreactors (TIBs) using Alka Burst technology addressed this issue and produced leafy biomass that expressed *OspA* at levels of up to 7.6% total soluble protein to give a maximum yield of *OspA*. This had proved to be nontoxic for plants and particularly useful when absolute gene dispersion control is a requisite [114]. Various recombinant therapeutic proteins have been produced from a single plant *C. reinhardtii*. Guy's 13 is a monoclonal antibody specific for a surface antigen of *Streptococcus mutans*, a bacterium that leads to dental caries. Chloroplast targets genetic construct having Guy's 13 gene under the transcriptional control of *psb*A promoter. This strategy was successful in expressing IgA-G, a humanized chimeric form of Guy's 13 antibody, with uniformly folded disulfide bonds [115].

### 9.2.4.2 Production of industrial enzymes and biomaterials

Chloroplast genetic engineering offers best-suited solution for the synthesis of economically and industrially important biomaterials such as amino acids, enzymes and technical proteins for multiple purposes. In one such instance, transgenic tomatoes were engineered to express milk proteins, human β casein, through CTT. Another important aspect is to produce biodegradable plastics that can be a miraculous alternative to synthetic plastics. Direct conversion of chorismate to *p*-hydroxybenzoic acid (*p*HBA) was achieved by integrating gene for chorismate pyruvate lyase (CPL), an enzyme encoded by the *ubiC* gene, in *E. coli*. A 10–20-fold increase in *p*HBA production in tobacco's nuclear transformation was noticed, which is far more than the required quantity for a commercially viable pathway [89]. However, the highest level of the poly (*p*HBA) polymer (25% dry weight) in normal healthy plants was optimized in chloroplast genomes [125]. An operon extension strategy was used for the production of the renewable biodegradable plastic polyhydroxy butyrate (PHB) in tobacco (*N. tabacum*) [126]. A lot of inputs have been witnessed to produce PHB through different systems, but till date maximum level of PHB was achieved in plastids only. This was due to the high substrate flux of the PHB pathway, that is, acetyl-CoA, during fatty acid biosynthesis in the plastids [127].

Xylanase is an important cellulolytic enzyme and biomaterial that finds its utility in the fiber, paper, animal feed and brewing industries. The first of nuclear transformation was unsuccessful in plants expressing xylanases due to cell wall degradation. However, 6% tsp xylanase accumulation was achieved when the *xynA* gene was integrated into the tobacco chloroplast genome [128].

Certain enzymes of amino acid biosynthesis are also targeted through CTT. One of the enzymes of tryptophan (Trp) synthesis and responsible for its regulation is anthranilate synthase (AS). Scientists engineered transgenic plants showing high levels of ASA2 mRNA. Higher expression of the AS α-subunit and a fourfold increase in AS enzyme activity were achieved, which was also found to be less sensitive to feedback inhibition of Trp biosynthesis. Resultant plants exhibited sevenfold increase in free Trp in the leaves of transgenic plant [129]. This finding has also opened new vistas of using *ASA2* gene of tobacco as a selectable marker. Economically important examples of enzymes and biomaterials that have been processed through chloroplast engineering within the tobacco genome are mentioned in the Table 9.5.

Table 9.5: Biomaterials and enzymes engineered via chloroplast genome of tobacco.

| S. No. | Biomaterials/enzymes | Gene | Expression level | Reference |
|---|---|---|---|---|
| 1 | Polyhydroxybutyrate | *phb operon* | 18.8% TSP | [93] |
| 2 | Xylanase | *xynA* | 6% TSP | [128] |
| 3 | β-Glucosidase | *Bgl1* | 20 mg/g TSP | [128] |
| 4 | Superoxide dismutase | *Cu/ZnSOD* | 9% TSP | [125] |
| 5 | Cellulases | *bgl1C, cel6B, cel9A,xeg74* | 5–40% TSP | [130] |
| 6 | *endo*-Glucanase | *celB* | 60–70% TSP | [131] |

### 9.2.4.3 Phytoremediation

Phytoremediation is a key solution to increasing threat of environmental pollution due to increasing urbanization and land uses. It is a safe and economic way for cleaning up environmental contaminants using plant system. Although plants possess limited potential to eradicate high levels of toxic chemical agents, their remediation capacity can be enhanced through genetic engineering aiming especially chloroplast genomes of plant. The most hazardous are the organomercurial compounds that mainly target chloroplast in plants. It is one of the most toxic forms of mercury. It is, thus, necessary to engineer chloroplast genome to provide added resistance and potentiate its detoxification ability for organomercurial compounds and metallic mercury [124,132]. Synthesis of metal chelators via CTT can do wonders in this issue. This will improve the capability of plants for metal uptake [7, 133, 134] and its sequestration. In this direction, Occhialini et al. integrated two bacterial genes encoding, mercuric ion reductase (*merA*) and organomercurial lyase (*merB*), as an operon in the chloroplasts of tobacco. Resultant transgenic plants accumulated mercury primarily in roots, reducing its concentration in soil, up to 200 μg/g. Furthermore, they are found to accumulate 100-fold more of mercury in the leaves than the control untransformed plants without showing any detrimental effect [135]. Phytoremediation of toxic mercury was also achieved by integrating metallothione in tobacco's chloroplast via CTT [136].

#### 9.2.4.4 Production of biofuels

CTT can also be explored in terms of production of biofuels. The foremost requisite for producing lingocellulose-based biofuels is to develop a rapid and cost-effective enzymatic system for speedy biomass depolymerization. Chloroplast acts as an ideal bioreactor for large-scale enzyme production due to its increased levels of protein expression and compartmentalization of toxic proteins within chloroplasts of transgenic plants [137]. Production of chloroplast-originated mixture of enzymes for production of fermentable sugars from various lignocellulosic biomass marked the stepping-stone in the area of biofuels. Biofuel-based researches are flooded with many examples of enzymes from various sources like bacteria, fungi and moulds, which are integrated into tobacco chloroplasts for the production of fermentable sugars. Few such examples are β-glucosidase, *β-1, 4-endoglucanase*, swollenin, cutinase, endoglucanases, esterase, exoglucanase, pectate lyases, acetyl xylan esterase, xylanase, xylan and lipase [61, 95, 138–140]. CTT-derived cutinase, due to its expansion activity, showed enlarged segments and easier unwinding of the intertwined inner fiber of cotton. It also exhibited esterase and lipase activity [95]. Another instance of β-mannanase enzyme isolated from *Trichoderma reesei* showed six- to sevenfold higher enzyme activity than *E. coli.* Pantaleoni et al. showed that when chloroplast-derived β-mannanase enzymatic mixture was used, 20% more glucose equivalents were produced from pinewood than the controls without mannanase [139]. Many other chloroplast-based enzymes like cutinase and pectate lyase A from *Fusarium solani*; endoglucanase exoglucanase from *Clostridium thermocellum*; endoglucanase, swollenin, acetyl xylan esterase and xylanase enzymes from *T. reesei* and lipase from

*Mycobacterium tuberculosis* showed broader pH and higher temperature stability than enzymes expressed in *E. coli*. In one of the studies, 36-fold higher glucose was produced from citrus peel; filter paper/pine wood when treated with chloroplast derived crude extract enzymatic cocktails as compared to the commercially available one [41].

#### 9.2.4.5 Conferring stress tolerance

As already mentioned, chloroplast is one of the best-suited sites for the improvement of many agronomic traits due to its higher level of gene expression and subsequent containment via maternal inheritance. Recently, conferring stress tolerance research area is gaining momentum. Many scientific groups are constantly working on targeting this miraculous organelle with special focus on herbicide resistance, insect resistance and tolerance to other biotic and abiotic stresses.

**Herbicide resistance**

Herbicide is the most studied and experimented area of the biotic stresses. Normally, transgenic plants developed for this trait are mostly insensitive to the hazardous effects of herbicides. Glyphosate is a broad-spectrum and most popular herbicide whose function is to inhibit the enzyme 5-enolpyruvylshikimate-3-phosphate (EPSPS) of shikimic acid pathway forming aromatic amino acids like tyrosine. It is encoded by a nuclear genome but is expressed within the chloroplast. Therefore, CTT plays a vital role in achieving glyphosate resistance [72, 73, 141]. The first successful report was of Daniell et al. [51]. They were able to express chloroplast-targeted eukaryotic gene in *petunia*. The *aroA* gene was inserted either between *trn*I and *trn*A genes in the inverted repeat region or between the *rbc*L and *acc*D genes in the large single copy region [51]. This stable integration was able to increase resistance even more than the lethal concentration. Another experiment showed the incorporation of *CP4 EPSPS* gene in the plastids of tobacco, which provided around 250 times more EPSPS enzyme in comparison to the nuclear transgenic experimentation [67]. The second most common herbicide used worldwide is phosphinothricin, commonly available as Basta/Bialaphos. Chloroplast engineering in tobacco plants by *bar* genes showed field-level tolerance when compared with the controls [82].

**Insect resistance**

Worldwide setbacks are incurred due to insect attack directly leading to production losses. Most of the strategies have aligned for insecticidal protoxins integration in plants produced by *B. thuringiensis*. The Cry2Aa2 is an operon, possessing target genes *Cry2Aa2*, *orf* 1 and *orf* 2. De Cosa et al. [43] recorded highest level of the chaperone activity of *orf* 2 and were able to express Cry2Aa2 protein up to 46% tsp in tobacco chloroplasts. The expressed foreign gene provided protection from the degradation by overexpressing chaperone crystal protein of *B. thuringiensis*. Reduced level of the same via the nuclear genome transformation poses the threat of developing Bt-resistant pests.

In contrast, CTT-derived plants were able to cause damage to the insects that could bear insecticidal protein concentrations as high as 40,000 times than normal, even at low levels of Cry2Aa2 gene expression [79]. In another set of experiment, Cry1Ia5 insecticidal protein was found to be accumulating up to the levels of 3% of tsp in tobacco leaf tissue by using protoplast transformation [142]. The antimicrobial peptide MSI-99 (an analogue of maganin-2) confers protection by targeting negatively charged phospholipids, which are especially found in the outer membranes of bacteria and fungi, against many prokaryotic organisms [143]. Additionally, fungal pathogen like *Colletotrichum* also showed similar responses by using chloroplast engineering.

The case study of MSI-99 in tobacco clearly indicated the contrasting features of chloroplast genome engineering over nuclear genome in conferring adequate disease resistance against pathogens. CTT was sufficient enough in providing protection from both bacterial and fungal pathogens [144].

#### 9.6.5.3 Osmotic stress resistance

Environmental stresses such as salinity, freezing and drought can affect extremely leading to osmotic imbalance and in turn cellular dehydration that restricts plant growth and development. Osmoprotectants are the choice of molecules whose function is to maintain osmotic balance and help the organism to survive under stressed conditions [145]. Solutes such as sugars, betaine and amino acids are produced in lieu of stressed condition. Their role is to support turgor pressure and maintain the integrity of cellular macromolecules and biological membranes from damage due to high salt concentrations. Most of the work in this direction is on trehalose, an osmoprotectant synthesized by trehalose phosphate synthase enzyme encoded by the TPS1 gene. Under stress conditions of heat, freezing, drought and salt, it accumulates and protects the plants from the deleterious effects. CTT-derived transgenics showed normal growth and were able to accumulate trehalose as much as 25 times higher [44].

Drought tolerance was analyzed by keeping CTT-derived transgenic plants in 6% polyethylene glycol (PEG) normal and healthy growth was observed in comparison to controls that were fairly bleached. The aforementioned threat of stress increases salinity in the soil, which is also due to excessive use of fertilizers and rainfed mode of irrigation in the present day. This not only reduces metabolite exchange but affects photosynthetic capacity in plants as well. Increasing the production of glycine betaine (GB) is found to be highly effective in many flowering plants and marine algae under such conditions. It is a quaternary ammonium osmolyte and hydrophilic in nature. However, many important cash crops do not synthesize or do not accumulate sufficient levels of GB. An experiment targeting the nuclear genome of spinach with *badh* (*betaine aldehyde dehydrogenase*) gene showed salt tolerance to a lesser extent. However, when carrot plastids were promoted to overexpress *badh* gene, a significant amount of tolerance was conferred in resultants due to the accumulation of GB against salt stress [80].

Furthermore, choline oxidase enzyme isolated from bacteria *Arthrobacter globiformis* was able to form GB in a single enzymatic reaction in spite of two in the former case. *Oryza sativa* L., which does not synthesize GB at all, when transformed with *coda* gene from *A. globiformis* via CTT showed stress tolerance to a greater extent. Another active osmolyte is an amino acid, that is, β-alanine. Increased expression of this in plants showed promising results in enhancing environmental stress tolerance. Decarboxylation of aspartate by enzyme l-aspartate-alpha-decarboxylase (AspDC) to form β-alanine and carbon dioxide in *E. coli* is under the control of *panD* gene encoding for AspDC. A constitutive *E. coli panD* expression cassette was coinserted into the chloroplast genome of tobacco with a constitutive *aadA* gene as a selectable marker gene through biolistic gene gun. This results in the accumulation of β-alanine in transplastomic plants four times higher than nontransgenics. The resultants also showed 30–40% increased biomass due to $CO_2$ assimilation in transplastomic plants under heat stress [83].

#### 9.2.4.6 Enhancing nutrition

Vitamin E-enriched oils are obtained from the seeds of rapeseed, soybean, groundnut and maize. The most abundant form of vitamin E is γ-tocopherol, but its biological active form is marked by α-tocopherol, which is relatively low in content. However, sunflower (*Helianthus annuus*) seed oil is quite rich in α-tocopherol [146]. α-Tocopherol is synthesized from its immediate precursor γ-tocopherol with the help of enzyme γ-tocopherol methyl transferase (γ-TMT). This also represents the rate-limiting step of the pathway [147].

Integration of the *γ-tmt* gene into the chloroplast genome of the seeds showed overexpression of γ-TMT in the inner layer of chloroplast. The conversion rate of γ-tocopherol to α-tocopherol was found to be approximately 10 times higher in seeds [81]. Synthesis of provitamin A, that is, β-carotene, is another example where lycopene β-cyclase (*lcy*) genes is integrated into the tomato plastids. This has enhanced conversion of lycopene into β-carotene many folds [148].

#### 9.2.4.7 Engineering nitrogen fixation

Nitrogen fixation is a costly affair in terms of energy utilization, that is, ATP consumption, and chloroplasts play a major role in bearing this energetic load through photosynthetic energy transduction products as a source for the same [149, 150].

Nitrogenase reductase enzyme is sensitive to the presence of $O_2$, which is the major by-product of photosynthesis and thus appears to restrict $N_2$ fixation by chloroplasts. However, cyanobacteria perform the nitrogen fixing process simultaneous to oxygenic photosynthesis. These two contrasting but important processes are possible by applying division of labor rule. Cyanobacteria segregate these two processes where thick-walled heterocyst fixes the atmospheric nitrogen as they are potent enough in blocking the entry of inhibitory oxygen. Other than this, heterocysts photosynthetic apparatus lacks PSII apparatus that produces oxygen as a product but at the same time contained PSI to proceed for cyclic photophosphorylation to gain ATP.

#### 9.2.4.8 Engineering of photosynthesis

Photosynthesis probes either by using chloroplast gene deletion or by mutagenesis. Bock and Khan [151], along with other groups, have tried to determine which ORFs of the chloroplast are vital for photosynthesis. Rubisco is the foremost choice of the scientists as a candidate for genetic engineering because of its potential for efficient atmospheric carbon fixation in vascular plants [152], primarily in C3 plants that lag behind C4 plants in terms of carbon-concentrating mechanisms [153, 154]. Gene replacement approaches have been explored to identify the factors that are responsible for the functioning of Rubisco like its preferences for $CO_2$ versus $O_2$ [155]. Kanevski and Maliga [156] found that functional Rubisco is formed when its nuclear-encoded large subunit is synthesized in the cytoplasm and with the aid of chloroplast transit peptide (CTP) is imported into chloroplasts. Fused C-terminus of the tobacco *rbcL* gene with a 6×Histag could also ascertain origin of functional Rubisco [157]. Transformation of heterologous Rubisco subunits *rbcL* from tomato of the same family into the chloroplast genome of tobacco revealed mixed results (Table 9.6). Hybrid enzymes with tobacco and tomato larger subunits produced pale/yellow-green plants that could grow photoautotrophically without the need of excess of $CO_2$ [158]. When the tobacco's larger subunit was exchanged with that of *Synechococcus* PCC6301, pale-yellow-colored plants were developed that needed higher concentration of sucrose for growth [159].

Results pertaining to the concentration of Rubisco needed for optimal growth and development of plants resulted in varied conclusions because of its sluggish nature. This is the reason that most of the part of stroma is filled with the same enzyme. Similar reason was proved in case of wheat [160]. However, results were contrasting in case of rice where addition of *rbcS* genes could not show much improvement in the photosynthesis [161]. In spite of this many plants showed normal phenotypes in greenhouse and in growth chambers under lower Rubisco levels. In another study, a thermostable Rubisco activase mutant was identified whose presence showed improved photosynthesis and growth of *Arabidopsis* [162], but to a limited extent as overexpression of the same was not beneficial in rice [163].

Table 9.6: Photosynthesis improvement through chloroplast transgenic expression and increased expression of various enzymes.

| S. No. | Type of gene alterations | References |
|---|---|---|
| 1 | Deletion of *rbcL* from plastome | [156] |
| 2 | Addition of 6×His onto *rbcL* in plastome | [157] |
| 3 | Replace tobacco *rbcL* with tomato *rbcL* | [158] |
| 4 | Replace tobacco *rbcL* with *rbcM* from *Rhodospirillum rubrum* | [164] |
| 5 | Replace tobacco *rbcL* with *rbcL* from C3 *Flaveria* | [165] |
| 6 | Replace tobacco *rbcL* with gene encoding-linked *Synechococcus* large and small subunits | [166] |
| 7 | Enzyme improvement<br>– Sedopheptalose-1,7-bisophosphatase (SBPase)<br>– Fructose-bisphosphate aldolase (FBP aldolase)<br>– Bicarbonate transporters<br>– Rubisco activase<br>– Bacterial glycolate to glycerate pathway (five proteins)<br>– Glycolate oxidase (GO), malate synthase (MS) [172]<br>– Cyanobacterial *ictB* gene [173] | [167, 168]<br>[167, 168]<br><br>[169]<br>[170]<br>[171]<br>[162] |

This has also been revealed that if one alters the concentration of enzymes of photosynthetic carbon assimilation and metabolism, expression of enzymes like SBPase would be increased and could possibly accelerate the rate of light-saturated photosynthetic reactions (Table 9.6). Examples of such candidate enzymes on which scientists have put in efforts are fructose-1, 6-bisphosphate aldolase (FBP aldolase), UDP-glucose phosphorylase and ADP-glucose pyrophosphorylase (ADPGPP) [174]. People have worked on improving photosynthetic efficiency by attempting to lower down the extent of photorespiration through CTT [162]. One major breakthrough was to get a C3 plant to behave like a C4 by introducing C4 photosynthetic machinery into C3 plants like famous C4 rice project [175, 176].

## 9.3 Limitations to plastid transformation technology

One of the leading constraints limiting CTT potential is the present inability to derive regenerated homoplasmic monocotyledonous transplastomic plants, especially in case of promoting photosynthesis in crop plants [177]. The case of monocotyledonous grasses demands a whole new range of selectable markers and plant tissue culture techniques to get the stable transformants. Possible choices feasible for CTT are flooded in many review articles [178, 179]. Major effort of the hour is to make nuclear transformation feasible for major crop plants where CTT is not finding its way. It is, however, necessary to widen up applications of CTT for a number of important plants.

## 9.4 Conclusion and future prospects

Enormous advantages are listed here of chloroplast genome over nuclear genome as far as plant genetic transformation efforts are concerned. The nuclear genome targeting strategies are not able to deliver products when a higher rate of expression and multigene engineering (more than one gene transfer engineering) are a prerequisite. Chloroplast transformation offers a wide range of unique advantages over other biotechnological applications as discussed in this chapter: improving photosynthetic efficacy, conferring agronomic traits, producing industrial important enzymes, biofuels, biomaterials, phytoremediation, molecular farming for antibiotics, biopharmaceuticals and vaccines. Many crops acted as a candidate for CTT such as tomato, tobacco, *Arabidopsis*, carrot, lettuce, potato, oilseed rape, sugar beet, cotton, petunia, cabbage, soybean, rice, poplar eggplant and sugarcane [180].

Many successful attempts have been made in CTT, but there is still long way ahead that might be more challenging than earlier cases as with the monocot grasses. Challenges are like recalcitrant nature of cereal crops as mentioned above to the available regeneration protocols. Obtaining an efficient stable shoot regeneration system is the key to success in these crops [181]. Second, standardization of huge level of expression foreign proteins is to be under control as the phenotypic alterations of transplastomic plants can lead to severe losses [182].

Other challenges are to tackle the lack of appropriate tissue-specific regulatory and promoter sequences [183, 184] and the case gene expression in nongreen plastids [185]. Efficient recovery of the desired transplastomic transformants [186] and degradation of foreign proteins includes other such hindrances to the CTT [187]. Moreover, CTT provides marker-free transplastomic plants that will certainly ease public concern and boost up the public acceptance for this miraculous organelle. Despite many constraints, chloroplast genome targeting is still an attractive site and gaining momentum for wonders to come, which will leave huge impact on humans and rest of the living world.

## References


[1] Hamblin, A. How do we know when agricultural systems are sustainable? In: Environmental Indicators for Sustainable Agriculture. Report on a national workshop, 28-29 November 1991, ed. Hamblin A. Bureau of Rural Resources, Land and Water Resource Research and Development Corporation, Grains Res. Corporation, Canberra, Australia, 1992, 90.

[2] http://www.fao.org

[3] Horsch RB, Fry JE, Hoffman NL, Eichholtz D, Rogers SG, Fraley RT. A simple and general method of transferring genes into plants. Science 1985, 227, 1229–31.

[4] Roa-Rodriguez C, Nottenburg C. *Agrobacterium*-mediated transformation of plants. Cambia technology landscape paper, 2003 (http://www.bios.net/Agrobacterium).

[5] Jones HD. Wheat transformation: current technology and applications to grain development and composition. J Cereal Sci 2005, 41, 137–47.

[6] Travella S, Ross SM, Harden J, Everett C, Snape JW, Harwood WA. A comparison of transgenic barley lines produced by particle bombardment and *Agrobacterium*-mediated techniques. Plant Cell Rep 2005, 23,780–9.

[7] Hussain SS, Husnain T, Riazuddin S. Sonicated assisted *Agrobacterium* mediated transformation (SAAT): an alternative method for cotton transformation. Pak J Bot 2007, 39, 223–30.

[8] Beranová M, Rakouský S, Vávrová Z, Skalický T. Sonication assisted *Agrobacterium* mediated transformation enhances the transformation efficiency in flax (*Linum usitatissimum* L.). Plant Cell Tiss Org Cult 2008, 94, 253–9.

[9] Johnston SA, Anziano PQ, Shark K, Sanford JC, Butow RA. Mitochondrial transformation in yeast by bombardment with microprojectiles. Science 1988, 240, 1538–41.

[10] Svab Z, Hajdukievitz P, Maliga P. Stable transformation of plastids in higher plants. Proc Natl Acad Sci USA 1990, 87, 8526–30.

[11] Christou P. Transformation technology. Trends in Plant Science 1996, 1, 423–31.

[12] Hansen G, Chilton MD. 'Agrolistic' transformation of plant cells: integration of T-strands generated in planta. Proc Natl Acad Sci USA 1996, 93, 14978–83.

[13] Hansen G, Shillito RD, Chilton MD. T-strand integration in maize protoplasts after co-delivery of a T-DNA substrate and virulence genes. Proc Nat Acad Sci USA 1997, 94, 11726–30.

[14] Paszkowski J, Shillito RD, Saul M, Mandak V, Hohn T, Hohn B, *et al.* Direct gene transfer to plants. EMBO J 1984, 3, 2717–22.

[15] Fromm M, Taylor LP, Walbot V. Expression of genes transferred into monocot and dicot plant cells by electroporation. Proc Natl Acad Sci USA 1985, 82, 5824–8.

[16] Weaver JC. Electroporation theory: concepts and mechanisms. In: Nickoloff JA, ed. Electroporation Protocols for Microorganisms. Totowa, New Jersey: Humana Press 1995, 1–26.

[17] Abdul-Baki AA, Saunders JA, Matthews BF, Pittarelli GW. DNA uptake during electroporation of germinating pollen grains. Plant Sci 1990, 70, 181–90.

[18] Morikawa H, Yamada Y. Capillary microinjection into protoplasts and intranuclear localization of injected materials. Plant Cell Physiol 1985, 26, 229–36.

[19] Lörz H, Paszkowski J, Dierks-Ventling C, Potrykus I. Isolation and characterization of cytoplasts and miniprotoplasts derived from protoplasts of cultured cells. Physiol Plant 1981, 53, 385–91.

[20] Zhu Z, Sun B, Liu C, Xiao G, Li X. Transformation of wheat protoplasts mediated by cationic liposome and regeneration of transgenic plantlets. Chin J Biotechnol 1993, 9, 257–61.

[21] Patnaik D, Khurana P. Wheat biotechnology—a minireview. Electr J Biotech 2001, 4, 74–102.

[22] Hans-Hening H, Hupfer B, Koch H, Ringsdorf H. Polymerizable phospholipids analogues—new stable biomembrane and cell models. Angew Chem Int Ed Engl 1980, 19, 938–40.

[23] Luo Z, Ray W. A simple method for the transformation of rice *via* the pollen-tube pathway. Plant Mol Biol Rep 1988, 6, 165–74.
[24] Mu HM, Liu SJ, Zhou WJ, Wen YX, Zhang WJ, Wei RX. Transformation of wheat with insecticide gene of arrow head proteinase inhibitors by pollen tube pathway and analysis of transgenic plants. I Chuan Hsueh Pao (Chinese) 2002, 26, 634–42.
[25] Shou HX, Palmer RG, Wang K. Irreproducibility of the soybean pollen-tube pathway transformation procedure. Plant Mol Biol Report 2002, 20, 325–34.
[26] Kaeppler HF, Somers DA, Rines HW, Cockburn AF. Silicon carbide fiber-mediated stable transformation of plant cells. Theoret Appl Genet 1992, 84, 560–6.
[27] Frame BR, Drayton PR, Bagnall SV, Lewnau CJ, Bullock P, Wilson HM, *et al.* Production of fertile transgenic maize plants by silicon carbide whisker mediated transformation. Plant J 1994, 6, 941–8.
[28] Dalton SJ, Bettany AJE, Timms E, Morris P. Transgenic plants of *Lolium multiflorum, Lolium perenne, Festuca arundinacea* and *Agrostis stolonifera* by silicon carbide fibre-mediated transformation of cell suspension cultures. Plant Sci 1998, 132, 31–43.
[29] Prabha S, Zhou WZ, Labhasetwar V. Particle size dependency of nanoparticle mediated gene transfection: Effect of particle size. Int J Pharm 2002, 244, 105–15.
[30] Zhu X, Lu L, Currier BL, Windebank AJ, Yaszemski MJ. Controlled release of NFκB decoy oligonucleotides from biodegradable polymer microparticles. Biomaterials 2002, 23, 2683–92.
[31] Gill JS, Zhu X, Moore MJ, Lu L, Yaszemski MJ, Windebank AJ. Effects of NFκB decoy oligonucleotides released from biodegradable polymer microparticles on a glioblastoma cell line. Biomaterials 2002, 23, 2773–81.
[32] Ruhlman T, Ahangari R, Devine A *et al.* Expression of cholera toxin B-proinsulin fusion protein in lettuce and tobacco chloroplasts-oral administration protects against development of insulitis in non-obese diabetic mice. Plant Biotechnol J 2007, 5, 495–510.
[33] Bock R. Plastid biotechnology: prospects for herbicide and insect resistance, metabolic engineering, and molecular farming. Curr Opin Biotechnol 2007, 18, 100–6.
[34] Bock R, Warzecha H. Solar-powered factories for new vaccines and antibiotics. Trends Biotechnol 2010, 28, 246–52.
[35] Clarke JL, Daniell H. Plastid biotechnology for crop production: present status and future perspectives. Plant Mol Biol 2011, 76, 211–20.
[36] Daniell H, Singh ND, Mason H, Streat field SJ. Plant-made vaccine antigens and biopharmaceuticals. Trends Plant Sci 2009, 14, 669–79.
[37] Daniell H, Choun-S L, Ming Y, Wan JC. Chloroplast genomes: diversity, evolution, and applications in genetic engineering. Genome Biol 2016, 17, 134.
[38] Heifetz PB. Genetic engineering of the chloroplast. Biochimie 2000, 82, 655–66.
[39] Maul JE, Lilly JW, Cui L, Depamphilis CW, Miller W, Harris EH, *et al.* The *Chlamydomonas reinhardtii* plastid chromosome: islands of genes in a sea of repeats. Plant Cell 2002, 14, 2659–79.
[40] Maliga P. Progress towards commercialization of plastid transformation technology. Trends Biotechnol 2003, 21(1),20–8.
[41] Ahmad N, Burgess SJ, Nielsen BL. Editorial: advances in plastid biology and its applications. Front Plant Sci 2016, 7, 1396.
[42] Sugiura M, Takeda Y. Nucleic acids. In: Buchanan B.B., Gruissem W., Jones R.L. (eds): Biochemistry and Molecular Biology of Plants, American Society of Plant Physiologists, Rockville, 2000, 260–310.
[43] De Cosa B, Moar W, Lee SB, Miller M, Daniell H. Over expression of the *Bt cry2Aa2* operon in chloroplasts leads to formation of insecticidal crystals. Nat Biotechnol 2001,19, 71–4.

[44] Lee SB, Kwon HB, Kwon SJ, Park SC, Jeong MJ, Han SE, Daniell H. Accumulation of trehalose within transgenic chloroplasts confers drought tolerance. Mol Breeding 2003, 11, 1–13.
[45] Daniell H, Lee SB, Panchal T, Wiebe PO. Expression of cholera toxin B subunit gene and assembly as functional oligomers in transgenic tobacco chloroplasts. J of Mol Biol 2001, 311, 1001–9.
[46] Nagata N, Saito C, Sakai A, Kuroiwa H, Kuroiwa T. The selective increase or decrease of organellar DNA in generative cells just after pollen mitosis one controls cytoplasmic inheritance. Planta 1999, 209, 53–65.
[47] Boynton JE, Gillham NW, Harris EH, Hosler JP, Johnson AM, Jones AR, *et al.* Chloroplast transformation in *Chlamydomonas* with high velocity microprojectiles. Sciences 1998, 240, 1534–8.
[48] Klein TM, Wolf ED, Wu R, Sanford JC. High velocity microprojectiles for delivering nucleic acids into living cells. Nature 1987, 327, 70–3.
[49] Finer JJ, Vain P, Jones MW, McMullen MD. Development of the particle inflow gun for DNA delivery to plant cells. Plant Cell Rep 1992, 11, 323–8.
[50] Jabeen R, Khan MS, Zafar Y, Anjum T. Codon optimization of *cry1Ab* gene for hyper expression in plant organelles. Mol Biol Rep 2010, 37, 1011–7.
[51] Daniell H, Datta R, Varma S, Gray S, Lee SB. Containment of herbicide resistance through genetic engineering of the chloroplast genome. Nat Biotechnol 1998, 16, 345–8.
[52] McBride KE, Svab Z, Schaaf DJ, *et al.* Amplification of a chimeric *Bacillus* gene in chloroplasts leads to an extraordinary level of an insecticidal protein in tobacco. Biotechnology 1995, 13, 362–5.
[53] O'Neill C, Horváth GV, Horváth E, Dix PJ, Medgyesy P. Chloroplast transformation in plants: polyethylene glycol (PEG) treatment of protoplasts is an alternative to biolistic delivery systems. Plant J 1993, 3, 729–38.
[54] Koop HU, Steinmüller K, Wagner H, Rößler C, Eibl C, Sacher L. Integration of foreign sequences into the tobacco plastome *via* polyethylene glycol-mediated protoplast transformation. Planta 1996, 199, 193–201.
[55] Knoblauch M, Hibberd JM, Gray JC, Aart JE van Bel. A galinstan expansion femtosyringe for microinjection of eukaryotic organelles and prokaryotes. Nat Biotechnol 1999, 17, 906–9.
[56] Hajdukiewicz PT, Allison LA, Maliga P. The two RNA polymerases encoded by the nuclear and the plastid compartments transcribe distinct groups of genes in tobacco plastids. EMBO J 1997, 16, 4041–8.
[57] Allison LA, Simon LD, Maliga P. Deletion of *rpoB* reveals a second distinct transcription system in plastids of higher plants. EMBO J 1996, 15, 2802–9.
[58] Muhlbauer SK, Koop HU. External control of transgene expression in tobacco plastids using the bacterial *lac* repressor. The Plant J 2005, 43, 941–6.
[59] Lössl A, Bohmert K, Harloff H, Eibl C, Muhlbauer S, Koop HU. Inducible trans-activation of plastid transgenes: expression of the *R. eutropha phb* operon in transplastomic tobacco. Plant Cell Physiol 2005, 46, 1462–71.
[60] Chakrabarti SK, Lutz KA, Lertwiriyawong B, Svab Z and Maliga P. Expression of the *cry9Aa2* Bt gene in tobacco chloroplasts confers resistance to potato tuber moth. Transgenic Res 2006, 15, 481–8.
[61] Gray BN, Yang H, Ahner BA, Hanson MR. An efficient downstream box fusion allows high-level accumulation of active bacterial beta-glucosidase in tobacco chloroplasts. Plant Mol Biol 2011, 76, 345–55.
[62] Herz S, Fussl M, Steiger S, Koop HU. Development of novel types of plastid transformation vectors and evaluation of factors controlling expression. Transgenic Res 2005, 14, 969–82.

[63] Barkan A. Expression of plastid genes: organelle-specific elaborations on a prokaryotic scaffold. Plant Physiol 2011, 155, 1520–32.
[64] Tregoning JS, Nixon P, Kuroda H, Svab Z, Clare S, Bowe F, Fairweather N, Ytterberg J, van Wijk KJ, Dougan G, Maliga P. Expression of tetanus toxin Fragment C in tobacco chloroplasts. Nucleic Acids Res 2003, 31, 1174–9.
[65] Yang H, Gray BN, Ahner BA, Hanson MR. Bacteriophage 5' untranslated regions for control of plastid transgene expression. Planta 2013, 237(2),517–27.
[66] Sprengart ML, Fuchs E, Porter AG. The downstream box: an efficient and independent translation initiation signal in *Escherichia coli*. EMBO J 1996, 15, 665–74.
[67] Ye GN, Hajdukiewicz PT, Broyles D, Rodriguez D, Xu CW, Nehra N, Staub JM. Plastid-expressed 5-enolpyruvylshikimate- 3-phosphate synthase genes provide high level glyphosate tolerance in tobacco. Plant J 2001, 25, 261–70.
[68] Gray BN, Ahner BA, Hanson MR. High-level bacterial cellulase accumulation in chloroplast-transformed tobacco mediated by downstream box fusions. Biotechnol and Bioengg 2009, 102, 1045–54.
[69] Stern DB, Goldschmidt-Clermont M, Hanson MR. Chloroplast RNA metabolism. Ann Rev of Plant Biol 2010, 61, 125–55.
[70] Dheeraj V, Daniell H. Chloroplast vector systems for biotechnology applications. Plant Physiol. 2007, 145, 1129.
[71] Chen PJ, Senthil kumar R, Jane WN, He Y, Tian Z, Yeh KW. Transplastomic *Nicotiana benthamiana* plants expressing multiple defence genes encoding protease inhibitors and chitinase display broad-spectrum resistance against insects, pathogens and abiotic stresses. Plant Biotechnol J 2014, 12, 503–15.
[72] Jana Ř. Potential of chloroplast genome in plant breeding. Czech J Genet Plant Breed 2010, 46(3),103–13.
[73] Haghani K, Salmanian AH, Ranjbar B, Zakikhan K, Khajeh K. Comparative studies of wild type *E. coli* 5-enolpyruvylshikimate 3-phosphate synthase with three glyphosate insensitive mutated forms, activity, stability and structural characterization. Biochem Biophys Acta 2008, 1784, 1167–75.
[74] Kang TJ, Loc NH, Jang MO, Jang YS, Kim YS, Seo JE, *et al.* Expression of the Beta subunit of *E. coli* heat-labile enterotoxin in the chloroplasts of plants and its characterization. Transgenic Res 2003, 12(6),683–91.
[75] Jin S, Zhang X, Daniell H. *Pinellia ternata* agglutinin expression in chloroplasts confers broad spectrum resistance against aphid, whitefly, Lepidopteran insects, bacterial and viral pathogens. Plant Biotechnol J. 2012, 10, 313–27.
[76] Jin S, Kanagaraj A, Verma D, Lange T, Daniell H. Release of hormones from conjugates: chloroplast expression of beta-glucosidase results in elevated phytohormone levels associated with significant increase in biomass and protection from aphids or whiteflies conferred by sucrose esters. Plant Physiol. 2011, 155, 222–35.
[77] Lee SB, Li B, Jin S, Daniell H. Expression and characterization of antimicrobial peptides Retrocyclin-101 and Protegrin-1 in chloroplasts to control viral and bacterial infections. Plant Biotechnol J. 2011, 9, 100–15.
[78] DeGray G, Rajasekaran K, Smith F, Sanford J, Daniell H. Expression of an antimicrobial peptide *via* the chloroplast genome to control phytopathogenic bacteria and fungi. Plant Physiol. 2001, 127, 852–62.
[79] Kota M, Daniell H, Varma S, Garczynski SF, Gould F, William M.J. Overexpression of the *Bacillus thuringiensis* (Bt) Cry2Aa2 protein in insects. Proc Natl Acad Sci USA 1999, 96, 1840–45.

[80] Kumar S, Dhingra A, Daniell H. Plastid-expressed betaine aldehyde dehydrogenase gene in carrot cultured cells, roots, and leaves confer enhanced salt tolerance. Plant Physiol 2004, 136, 2843–54.
[81] Jin S, Daniell H. Expression of gamma-tocopherol methyltransferase in chloroplasts results in massive proliferation of the inner envelope membrane and decreases susceptibility to salt and metal-induced oxidative stresses by reducing reactive oxygen species. Plant Biotechnol J 2014, 12, 1274–85.
[82] Lutz KA, Knapp JE, Maliga P. Expression of bar in the plastid genome confers herbicide resistance. Plant Physiol 2001, 125, 1585–90.
[83] Fouad WN, Altpeter F. Transplastomic expression of bacterial l-aspartate-alpha-decarboxylase enhances photosynthesis and biomass production in response to high temperature stress. Transgenic Res 2009,18, 707–18.
[84] Wurbs D, Ruf S, Bock R. Contained metabolic engineering in tomatoes by expression of carotenoid biosynthesis genes from the plastid genome. Plant J 2007, 49, 276–88.
[85] Lu Y, Rijzaani H, Karcher D, Ruf S, Bock R. Efficient metabolic pathway engineering in transgenic tobacco and tomato plastids with synthetic multigene operons. Proc Natl Acad Sci USA 2013, 110, E623–32.
[86] Dhingra A, Portis Jr AR, Daniell H. Enhanced translation of a chloroplast expressed RbcS gene restores small subunit levels and photosynthesis in nuclear RbcS antisense plants. Proc Natl Acad Sci USA 2004, 101, 6315–20.
[87] Yabuta Y, Tanaka H, Yoshimura S, Suzuki A, Tamoi M, Maruta T, Shigeoka S. Improvement of vitamin E quality and quantity in tobacco and lettuce by chloroplast genetic engineering. Transgenic Res. 2013, 22, 391–402.
[88] Pengelly JJ, Forster B, von Caemmerer S, Badger MR, Price GD, Whitney SM. Transplastomic integration of a cyanobacterial bicarbonate transporter into tobacco chloroplasts. J Exp Bot 2014, 65, 3071–80.
[89] Viitanen PV, Devine AL, Kahn S, Deuel DL, Van Dyk DE, Daniell H. Metabolic engineering of the chloroplast genome using the *E. coli ubiC* gene reveals that chorismate is a readily abundant precursor for 4-hydroxybenzoic acid synthesis in plants. Plant Physiol 2004, 136, 4048–60.
[90] Verma D, Jin S, Kanagaraj A, Singh ND, Daniel J, Kolattukudy PE, *et al.* Expression of fungal cutinase and swollenin in tobacco chloroplasts reveals novel enzyme functions and/or substrates. PLoS One. 2013, 8, e57187.
[91] Elghabi Z, Karcher D, Zhou F, Ruf S, Bock R. Optimization of the expression of the HIV fusion inhibitor cyanovirin-N from the tobacco plastid genome. Plant Biotechnol J 2011, 9, 599–608.
[92] Yu LX, Gray BN, Rutzke CJ, Walker LP, Wilson DB, Hanson MR. Expression of thermostable microbial cellulases in the chloroplasts of nicotine-free tobacco. J Biotechnol 2007, 131, 362–9.
[93] Petersen K, Bock R. High-level expression of a suite of thermostable cell wall-degrading enzymes from the chloroplast genome. Plant Mol Biol 2011, 76, 311–21.
[94] Hasunuma T, Miyazawa SI, Yoshimura S, Shinzaki Y, Tomizawa KI, Shindo K, *et al.* Biosynthesis of astaxanthin in tobacco leaves by transplastomic engineering. Plant J 2008, 55, 857–68.
[95] Nakahira Y, Ishikawa K, Tanaka K, Tozawa Y, Shiina T. Overproduction of hyperthermostable beta-1,4-endoglucanase from the archaeon *Pyrococcus horikoshii* by tobacco chloroplast engineering. Biosci Biotechnol Biochem 2013, 77, 2140–3.
[96] Sherman A, Su J, Lin S, Wang X, Herzog RW, Daniell H. Suppression of inhibitor formation against FVIII in a murine model of hemophilia A by oral delivery of antigens bioencapsulated in plant cells. Blood 2014, 124, 1659–68.
[97] Boyhan D, Daniell H. Low-cost production of proinsulin in tobacco and lettuce chloroplasts for injectable or oral delivery of functional insulin and C-peptide. Plant Biotechnol J 2011, 9, 585–98.

[98] Youm JW, Jeon JH, Kim H, Min SR, Kim MS, Joung H, *et al.* High-level expression of a human beta-site APP cleaving enzyme in transgenic tobacco chloroplasts and its immunogenicity in mice. Transgenic Res. 2010, 19, 1099–108.
[99] Arlen PA, Falconer R, Cherukumilli S, Cole A, Cole AM, Oishi KK, Daniell H. Field production and functional evaluation of chloroplast-derived interferon-α2b Plant Biotechnol J 2007, 5, 511–25.
[100] Leelavathi S, Reddy VS. Chloroplast expression of His-tagged GUS-fusions: a general strategy to overproduce and purify foreign proteins using transplastomic plants as bioreactors. Mol Breeding 2003, 11, 49–58.
[101] Molina A, Hervas-Stubbs S, Daniell H, Mingo-Castel AM, Veramendi J. High yield expression of a viral peptide animal vaccine in transgenic tobacco chloroplasts. Plant Biotechnol J 2004, 2, 141–53.
[102] Soria-Guerra RE, Alpuche-Solis AG, Rosales-Mendoza S, Moreno-Fierros L, Bendik EM, Martinez-Gonzalez L, Korban SS. Expression of a multi-epitope DPT fusion protein in transplastomic tobacco plants retains both antigenicity and immunogenicity of all three components of the functional oligomer. Planta 2009, 229, 1293–302.
[103] Zhou YX, Lee MYT, Ng JMH, Chye ML, Yip WK, Zee SY, Lam E. A truncated hepatitis E virus ORF2 protein expressed in tobacco plastids is immunogenic in mice. World J Gastroenterol. 2006, 12, 306–12.
[104] Gonzalez-Rabade N, McGowan EG, Zhou F, McCabe MS, Bock R, Dix PJ, *et al.* Immunogenicity of chloroplast-derived HIV-1 p24 and a p24-Nef fusion protein following subcutaneous and oral administration in mice. Plant Biotechnol J 2011, 9, 629–38.
[105] Farran I, McCarthy-S I, Río-M F, Mansilla C, Lasarte JJ, Mingo-CM. The vaccine adjuvant extra domain A from fibronectin retains its proinflammatory properties when expressed in tobacco chloroplasts. Planta 2010, 231, 977–90.
[106] Spök A, Karner S, Stein AJ, Rodríguez-C E. Plant molecular farming, opportunities and challenges. JRC Scientific and Tech Rep. 2008, 1–146.
[107] Holtz BR, Berquist BR, Bennett LD, Kommineni VJ, Munigunti RK, White EL. Commercial-scale biotherapeutics manufacturing facility for plant made pharmaceuticals. Plant Biotechnol J 2015, 13, 1180–90.
[108] Rosales-Mendoza S, Soria-Guerra RE, Moreno-Fierros L, Alpuche-Solís AG, Martínez-González L, Korban SS. Expression of an immunogenic F1-V fusion protein in lettuce as a plant-based vaccine against plague. Planta 2010, 232, 409–16.
[109] Almaraz-Delgado AL, Flores-Uribe J, Pérez-España VH, Salgado-Manjarrez E, Badillo-Corona JA. Production of therapeutic proteins in the chloroplast of *Chlamydomonas reinhardtii*. AMB Express 2014, 4, 57.
[110] Daniell H, Chebolu S, Kumar S, Singleton M, Falconer R. Chloroplast derived vaccine antigens and other therapeutic proteins. Vaccine 2005, 23, 1779–83.
[111] Davoodi-Semiromi A, Schreiber M, Nalapalli S, Verma D, Singh ND, Banks RK, *et al.* Chloroplast-derived vaccine antigens confer dual immunity against cholera and malaria by oral or injectable delivery. Plant Biotechnol J 2010, 8(2),223–42.
[112] Michoux F, Ahmad N, Hennig A, Nixon PJ, Warzecha H. Production of leafy biomass using temporary immersion bioreactors: an alternative platform to express proteins in transplastomic plants with drastic phenotypes. Planta 2013, 237, 903–8.
[113] Lakshmi PS, Verma D, Yang X, Lloyd B, Daniell H. Low cost tuberculosis vaccine antigens in capsules: expression in chloroplasts, bioencapsulation, stability and functional evaluation *in vitro*. PLoS ONE 2013, 8, e54708.
[114] Tran M, Van C, Barrera DJ, Pettersson PL, Peinado CD, Bui J, *et al.* Production of unique immunotoxin cancer therapeutics in algal chloroplasts. Proc Natl Acad Sci US 2013,110, E15–22.

[115] Daniell H, Cohill PR, Kumar S, Dufourmantel N. Chloroplast genetic engineering. In: Da niell H., Chase C.D. eds: Molecular Biology and Biotechnology of Plant Organelles. Kluwer Academic Publishers, Dordrecht, 2004, 443–90.
[116] Morgenfeld M, Lentz E, Segretin ME, Alfano EF, Bravo-Almonacid F. Translational fusion and redirection to thylakoid lumen as strategies to enhance accumulation of human papilloma virus E7 antigen in tobacco chloroplasts. Mol Biotechnol 2014, 56, 1021–31.
[117] Gorantala J, Grover S, Rahi A, Chaudhary P, Rajwanshi R, Sarin NB, *et al.* Generation of protective immune response against anthrax by oral immunization with protective antigen plant-based vaccine. J Biotechnol 2014, 176, 1–10.
[118] Oey M, Lohse M, Kreikemeyer B, Bock R. Exhaustion of the chloroplast protein synthesis capacity by massive expression of a highly stable protein antibiotic. Plant J 2009, 57, 436–45.
[119] Lim S, Ashida H, Watanabe R, Inai K, Kim Y-S, Mukougawa K, *et al.* Production of biologically active human thioredoxin 1 protein in lettuce chloroplasts. Plant Mol Biol 2011, 76, 335–44.
[120] Wang YP, Wei ZY, Zhong XF, Lin CJ, Cai YH, Ma J, *et al.* Stable expression of basic fibroblast growth factor in chloroplasts of tobacco. Int J Mol Sci 2016, 17, 9–18.
[121] Morgenfeld M, Segretin ME, Wirth S, Lentz E, Zelada A, Mentaberry A, *et al.* Potato virus X coat protein fusion to human papilloma virus16 E7 oncoprotein enhance antigen stability and accumulation in tobacco chloroplast. Mol Biotechnol 2009, 43(3),243–9.
[122] Gregory JA, Topol AB, Doerner DZ, Mayfield S. Alga-produced choleratoxin-Pfs25 fusion proteins as oral vaccines. Appl Environ Microbiol 2013, 79, 3917–25.
[123] Yoon S-M, Kim SY, Li KF, Yoon BH, Choe S, Kuo MM-C. Transgenic microalgae expressing *Escherichia coli* App Aphytase as feed additive to reduce phytate excretion in the manure of young broiler chicks. Appl Microbiol Biotechnol 2011, 91, 553–63.
[124] Kupper H, Kepper F, Spiller M. Environmental relevance of heavy metal substituted chlorophylls using the example of water plants. J Exp Bot 1996, 47, 259–66.
[125] Bohmert-Tatarev K, McAvoy S, Daughtry S, Peoples OP, Snell KD. High levels of bioplastic are produced in fertile transplastomic tobacco plants engineered with a synthetic operon for the production of polyhydroxybutyrate. Plant Physiol 2011, 155, 1690–708.
[126] Beilen JB, Poirier Y. Production of renewable polymers from crop plants. Plant J. 2008, 54, 684–701.
[127] Snell KD, Peoples OP. PHA bioplastic: a value-added coproduct for biomass biorefineries. Biofuel Bioprod Bior 2009, 3, 456–67.
[128] Guda C, Lee SB, Daniell H. Stable expression of a biodegradable protein based polymer in tobacco chloroplasts. Plant Cell Rep 2000, 19, 257–62.
[129] Barone P, Zhang XH, Widholm JM Tobacco plastid transformation using the feedback insensitive anthranilate synthase alpha-subunit of tobacco (*ASA2*) as a new selectable marker. J of Exp Bot 2009, 60, 3195–202.
[130] Espinoza-Sánchez EA, Álvarez-Hernández MH, Torres-Castillo JA, Rascón-Cruz Q, Gutiérrez-Díez A, Zavala-García F, *et al.* Stable expression and characterization of a fungal pectinase and bacterial peroxidase genes in tobacco chloroplast. Electron J Biotechnol 2016, 18, 161–8.
[131] Madanala R, Gupta V, Pandey AK, Srivastava S, Pandey V, Singh PK, *et al.* Tobacco chloroplasts as bioreactors for the production of recombinant superoxide dismutase in plants, an industrially useful enzyme. Plant Mol Biol Rep 2015, 33, 1107–15.
[132] Smits E, Pilon M. Phytoremediation of metals using transgenic plants. Crit Rev Plant Sci 2002, 21, 439–56.
[133] Clemens S, Palmgren MG, Kranmer U. A long way ahead: understanding and engineering plant metal accumulation. Trends Plant Sci 2002, 7, 309–14.
[134] Ruiz ON, Hussein H, Terry N, Daniell H. Phytoremediation of organomercurial compounds *via* chloroplast genetic engineering. Plant Physiology 2003, 132, 1344–52.

[135] Occhialini A, Lin MT, Andralojc PJ, Hanson MR, Parry MAJ. Transgenic tobacco plants with improved cyanobacterial Rubisco expression but no extra assembly factors grow at near wild-type rates if provided with elevated $CO_2$. Plant J 2016, 85, 148–60.
[136] Agrawal P, Verma D, Daniell H. Expression of *Trichoderma reesei* beta mannanase in tobacco chloroplasts and its utilization in lignocellulosic woody biomass hydrolysis. PLoS ONE 2011, 6, e29302.
[137] Jin S, Daniell H. Engineered chloroplast genome just got smarter. Trends Plant Sci 2015, 20 (10), 622–40.
[138] Kim JY, Kanagaraj A, Verma D, Lange T, Daniell H. Production of hyper thermostable GH10 xylanase Xyl10B from *Thermotoga maritime* in transplastomic plants enables complete hydrolysis of methyl glucuronoxylan to fermentable sugars for biofuel production. Plant Mol Biol 2011, 76, 357–69.
[139] Pantaleoni L, Longoni P, Ferroni L, Baldisserotto C, Leelavathi S, Reddy VS, Pancaldi S, *et al.* Chloroplast molecular farming: efficient production of a thermostable xylanase by *Nicotiana tabacum* plants and long-term conservation of the recombinant enzyme. Protoplasma 2014, 251, 639–48.
[140] Xue XY, Mao YB, Tao XY, Huang YP, Chen XY. New approaches to agricultural insect pest control based on RNA interference. Adv Insect Physiol 2012, 42, 73–117.
[141] Roudsari MF, Salmanian AH, Mousavi A, Sohi HH, Jafari M. Regeneration of glyphosate-tolerant *Nicotiana tabacum* after plastid transformation with a mutated variant of bacterial *aroA* gene. Iran J Biotechnol 2009, 7, 247–53.
[142] Reddy VS, Leelavanthi S, Selvapandiyan A, Raman R, Giovanni F, Shukla V *et al.* Analysis of chloroplast transformed tobacco plants with *cry1Ia5* under rice *psbA* transcription elements reveal high level expression of *Bt* toxin without imposing yield penalty and stable inheritance of transplastome. Mol Bred 2002, 9, 259–69.
[143] Houston M.E.Jr, Kondejevski L, Gough M, Fidai S, Hodges RS, Hancock R. Influence of performed α-helix and α-helix induction on the activity of cationic antimicrobial peptides. J of Peptide Res 1997, 52, 81–88.
[144] Li Q, Lawrence CB, Xing HY, Babbit RA, Bass WT, Maiti IB, Everett NP. Enhanced disease resistance conferred by expression of an antimicrobial magainin analog in transgenic tobacco. Planta 2001, 212, 635–9.
[145] Glick BR, Pasternak JJ. Molecular Biotechnology: Principles and Applications of Recombinant DNA. American Soc Microbio, Washington, DC 1998, 463.
[146] Schneider C. Chemistry and biology of vitamin E Mol Nutr Food Res 2005, 49, 7–30.
[147] Shintan, Della Penna. Elevating the vitamin E content of plants through metabolic engineering, Science 1998, 282, 2098–100.
[148] Apel W, Bock R. Enhancement of carotenoid biosynthesis in transplastomic tomatoes by induced lycopene-to-provitamin A conversion. Plant Physiol 2009, 151, 59–66.
[149] Bogorad L. The chloroplast, its genome and possibilities for genetically manipulating plants. In *Genetic Engineering* (Vol. 1) Eds. Setlow JK and Hollaender A, Plenum Press 1979, 181–203.
[150] Merrick M, Dixon R. Why don't plants fix nitrogen? Trends Biotechnol 1984, 2, 162–6.
[151] Bock R, Khan MS. Taming plastids for a green future. Trends in Biotech 2004, 22, 311–8.
[152] Long SP, Zhu XG, Naidu SL, Ort DR. Can improvement in photosynthesis increase crop yields? Plant, Cell & Env 2006, 29, 315–30.
[153] Kajala K, Covshoff S, Karki S, Woodfield H, Tolley BJ, Dionora MJ, *et al.* Strategies for engineering a two-celled C(4) photosynthetic pathway into rice. J of Exp Bot 2011, 62, 3001–10.
[154] Peterhansel C. Best practice procedures for the establishment of a C(4) cycle in transgenic C(3) plants. J of Exp Bot 2011, 62, 3011–19.

[155] Mueller-Cajar O, Whitney SM. Directing the evolution of Rubisco and Rubisco activase: first impressions of a new tool for photosynthesis research. Photosynthesis Res 2008, 98, 667–75.
[156] Kanevski I, Maliga P. Relocation of the plastid *rbcL* gene to the nucleus yields functional ribulose-1, 5-bisphosphate carboxylase in tobacco chloroplasts. Proc Nat Acad Sci USA 1994, 91, 1969–73.
[157] Rumeau D, Becuwe-Linka N, Beyly A, Carrier P, Cuine S, Genty B, *et al.* Increased zinc content in transplastomic tobacco plants expressing a polyhistidine-tagged Rubisco large subunit. Plant Biotech J 2004, 2, 389–99.
[158] Zhang XH, Webb J, Huang YH, Lin L, Tang RS, Liu A. Hybrid Rubisco of tomato large subunits and tobacco small subunits is functional in tobacco plants. Plant Sci 2011, 180, 480–8.
[159] Kanevski I, Maliga P, Rhoades DF, Gutteridge S. Plastome engineering of ribulose-1,5-bisphosphate carboxylase/oxygenase in tobacco to form a sunflower large subunit and tobacco small subunit hybrid. Plant Physiol 1999, 119, 133–42.
[160] Theobald JC, Mitchell RA, Parry MA, Lawlor DW. Estimating the excess investment in ribulose-1,5-bisphosphate carboxylase/oxygenase in leaves of spring wheat grown under elevated $CO_2$. Plant Physiol 1998, 118, 945–55.
[161] Suzuki Y, Ohkubo M, Hatakeyama H, Ohashi K, Yoshizawa R, Kojima S, *et al.* Increased Rubisco content in transgenic rice transformed with the 'sense' rbcSgene. Plant Cell Physiol 2007, 48, 626–37.
[162] Kebeish R, Niessen M, Thiruveedhi K, Bari R, Hirsch HJ, Rosenkranz R, *et al.* Chloroplastic photorespiratory bypass increases photosynthesis and biomass production in *Arabidopsis thaliana*. Nature Biotechnol 2007, 25, 593–9.
[163] Kurek I, Chang TK, Bertain SM, Madrigal A, Liu L, Lassner MW, Zhu G. Enhanced Thermostability of *Arabidopsis* Rubisco activase improves photosynthesis and growth rates under moderate heat stress. The Plant Cell 2007, 19, 3230–41.
[164] Whitney SM, Andrews TJ. Plastome-encoded bacterial ribulose-1,5-bisphosphate carboxylase/oxygenase (RubisCO) supports photosynthesis and growth in tobacco. Proc Nat Acad Sci USA 2001b, 98, 14738–43.
[165] Whitney SM, Sharwood RE, Orr D, White SJ, Alonso H, Galmes J. Isoleucine 309 acts as a C4 catalytic switch that increases ribulose-1,5-bisphosphate carboxylase/oxygenase (rubisco) carboxylation rate in *Flaveria*. Proc Nat Acad Sci USA 2011b, 108, 14688–93.
[166] Whitney SM, Kane HJ, Houtz RL, Sharwood RE. Rubisco oligomers composed of linked small and large subunits assemble in tobacco plastids and have higher affinities for $CO_2$ and $O_2$. Plant Physiol 2009, 149, 1887–95.
[167] Feng L, Wang K, Li Y, Tan Y, Kong J, Li H, Zhu Y. Overexpression of SBPase enhances photosynthesis against high temperature stress in transgenic rice plants. Plant Cell Rep 2007, 26, 1635–46.
[168] Rosenthal DM, Locke AM, Khozaei M, Raines CA, Long SP, Ort DR. Over-expressing the C(3) photosynthesis cycle enzyme Sedoheptulose-1–7 Bisphosphatase improves photosynthetic carbon gain and yield under fully open air CO(2) fumigation (FACE). BMC Plant Bio 2011, 11, 123.
[169] Uematsu K, Suzuki N, Iwamae T, Inui M and Yukawa H. Increased fructose 1,6-bisphosphate aldolase in plastids enhances growth and photosynthesis of tobacco plants. J of Exp Bot 2012, 63, 3001–9.
[170] Price GD, Badger MR, von Caemmerer S. The prospect of using cyanobacterial bicarbonate transporters to improve leaf photosynthesis in C3 crop plants. Plant Physiol 2011, 155, 20–6.
[171] Fukayama H, Ueguchi C, Nishikawa K, Katoh N, Ishikawa C, Masumoto C, Hatanaka T and Misoo S. Overexpression of rubisco activase decreases the photosynthetic $CO_2$ assimilation rate by reducing rubisco content in rice leaves. Plant and Cell Physiol 2012, 53, 976–86.

[172] Maier A, Fahnenstich H, von Caemmerer S, Engqvist MK, Weber AP, Flugge UI, *et al.* Transgenic introduction of a glycolate oxidative cycle into *A. thaliana* chloroplasts leads to growth improvement. Front in Plant Sci 2012, 3, 38.
[173] Lieman-Hurwitz J, Rachmilevitch S, Mittler R, Marcus Y and Kaplan A. Enhanced photosynthesis and growth of transgenic plants that express *ictB*, a gene involved in HCO3-accumulation in cyanobacteria. Plant Biotechnol J 2003, 1, 43–50.
[174] Zhu XG, de Sturler E, Long SP. Optimizing the distribution of resources between enzymes of carbon metabolism can dramatically increase photosynthetic rate: a numerical simulation using an evolutionary algorithm. Plant Physiol 2007, 145, 513–26.
[175] Covshoff S, Hibberd JM. Integrating C4 photosynthesis into C3 crops to increase yield potential. Cur Opinions in Biotech 2012, 23, 209–14.
[176] Ruan CJ, Shao HB, Teixeira da Silva JA. A critical review on the improvement of photosynthetic carbon assimilation in C3 plants using genetic engineering. Crit Rev in Biotech 2012, 32, 1–21.
[177] Khan MS. Engineering photorespiration in chloroplasts: a novel strategy for increasing biomass production. Trends in Biotech 2007, 25, 437–40.
[178] Cardi T, Lenzi P, Maliga P. Chloroplasts as expression platform for plant produced vaccines. Expert Rev Vaccines. 2010, 9, 893–911.
[179] Day A, Michel GC. The chloroplast transformation toolbox: selectable markers and marker removal. Plant Biotechnol J 2011, 9, 540–53.
[180] Ahmadabadi M, Ruf S, Bock R. A leaf-based regeneration and transformation system for maize (*Zea mays* L.). Transgenic Res 2007, 16, 437–48.
[181] Ahmad N, Michoux F, Nixon PJ. Investigating the production of foreign membrane proteins in tobacco chloroplasts: expression of an algal plastid terminal oxidase. PLoS ONE 2012, 7, 417–22.
[182] Kahlau S, Bock R. Plastid transcriptomics and translatomics of tomato fruit development and chloroplast-to-chromoplast differentiation: chromoplast gene expression largely serves the production of a single protein. Plant Cell 2008, 20, 856–74.
[183] Valkov T, Scotti N, Kahlau S, MacLean D, Grillo S, Gray JC, *et al.* Genome wide analysis of plastid gene expression in potato leaf chloroplasts and tuber amyloplasts: transcriptional and posttranscriptional control. Plant Physiol 2009, 150, 2030–44.
[184] Zhang J, Ruf S, Hasse C, Childs L, Scharff LB, Bock R. Identification of *cis*-elements conferring high levels of gene expression in non-green plastids. Plant J 2012, 72, 115–28.
[185] Iamtham S, Day A. Removal of antibiotic resistance genes from transgenic tobacco plastids. Nat Biotechnol 2000,18, 1172–6.
[186] Zhou F, Badillo-Corona JA, Karcher D, Gonzalez-Rabade N, Piepenburg K, Borchers AMI, *et al.* High-level expression of human immunodeficiency virus antigens from the tobacco and tomato plastid genomes. Plant Biotechnol J 2008, 6, 897–913.
[187] Apel W, Schulze WX, Bock R. Identification of protein stability determinants in chloroplasts. Plant J 2010, 63, 636–50.

Ram Kumar Pundir, Pranay Jain, Satish Kumar,
Mukesh Yadav, Rajesh Kumar

# 10 Plant biotechnology: Industrial prospects and scopes

**Abstract:** Plant biotechnology involves use of multiple techniques in a systematic and productive manner to accomplish the needs and objectives of humans along with associated lives. Plant biotechnology covers nearly all aspects, from molecular to product level, for production of value-added ecofriendly industrial products in a cost-effective manner for human welfare. Plant biotechnology as a subject has very broad industrial future perspectives for human welfare. Plants have a tremendous potential to produce metabolites, enzymes, medicines and edible vaccines and remediate the unwanted substances using recombinant DNA technology. Plants-derived vaccines, medicines and secondary metabolites are of immense industrial potential. Plants secondary metabolites are of particular interest in different industries, including food and pharmaceuticals. In the current scenario, the world population is more inclined toward the plant products as compared to their pure chemical alternatives. This chapter focuses on industrial applications and future prospective of plant biotechnology.

**Keywords:** commercial aspects, edible vaccines, enzymes, metabolites, phytomedicine, phytoremediation, plant biotechnology, value addition

## 10.1 Introduction

Plant biotechnology is an important tool of modern science to satisfy the human and associated lives. Plant biotechnology may be defined as generation of useful products and services from plant cells, tissues and various other parts of plants. Plant biotechnology can feed the increasing population and also support the other commercial requirements of humans at global level. Plant tissue culture and plant genetic engineering are among the most important applied aspects of plant biotechnology. Micropropagation involves the practice of rapidly multiplying stock plant material to produce a large number of progeny plants with the help of modern plant tissue culture methods. Micropropagation is used to multiply plants that have been genetically modified or bred through conventional plant breeding methods. Various plant-based products, including secondary metabolites, carotenoids, pigments, oils, resins, flowers, fruits, seeds, medicines, edible vaccines and other components, are continuously in demand. Some examples of plant metabolites and plant-derived products are listed in Table 10.1.

https://doi.org/10.1515/9783110563337-010

Table 10.1: Some examples of plants having potential to be explored at industrial level for various purposes.

| Category of product | Compounds/plant sources | Properties |
|---|---|---|
| Flavonoids | – Anthocyanins<br>– Chalcone<br>– Flavanones<br>– Flavones<br>– Flavonols<br>– Isoflavonoid | – Antioxidant<br>– Anti-inflammatory<br>– Anticarcinogenic<br>– Antimicrobial<br>– Antimutagenic<br>– Potent inhibitor of some enzymes |
| Phytomedicine | – *Allium sativum*<br>– *Plumbago indica*<br>– *Semecarpus anacardium*<br>– *Hemidesmus indicus*<br>– *Terminalia arjuna*<br>– *Withania somnifera*<br>– *Ocimum sanctum* | – Cardiovascular disease |
| Phytomedicine/herbal formulation | – *Emblica officinalis*<br>– *Terminalia chebula*<br>– *Terminalia bellirica* | – Arthritic disease |
| Phytomedicine/herbal formulation | – *Echinacea purpurea*<br>– *Echinacea pallida* | – Respiratory tract disease |
| Phytomedicine/herbal formulation | – *E. officinalis*<br>– *T. chebula*<br>– *T. belarica* | – Digestive system disease/disorders |
| Plant pigment as bio-color | – Chlorophyll<br>– Betacyanins<br>– Flavonoids<br>– Anthocyanins<br>– Carotenoids | – Green-yellow color<br>– red-orange-yellow, bluish-red, red-violet<br>– Red, purple<br>– Blue<br>– Yellow-orange-red |
| Phytoremediation | – *Cynodon dactylon*<br>– *Paspalum notatum*<br>– *Populus nigra*<br>– *Brassica juncea*<br>– *Brassica campestris*<br>– *Thlaspi caerulescens*<br>– *Arabidopsis* sp. | – Heavy metals<br>– Pollutants |
| Secondary metabolites | – Terpenes and steroids<br>– Phenolic compounds<br>– Polyketide and fatty acids<br>– Phenylpropanoids<br>– Alkaloids<br>– Flavonoids | – Health benefits<br>– Perfume industry<br>– Cosmetics<br>– Food supplements<br>– Bio-colors |

Using scientific tools and techniques, expressions of plant molecules, metabolites and traits can be controlled to make a desired level of changes. Moreover, with the help of plant biotechnology, the nutritional value of raw food can be increased by altering the specific metabolic pathway. Scientists are extensively working one expanding the commercial dimensions of plant biotechnology and the related fields. This chapter focuses on major plant-derived commercial aspects or value-added products of prime importance.

## 10.2 Plant secondary metabolites and phytomedicines

A large number of structurally diverse secondary metabolites occur in plants and are stored in higher concentrations, sometimes in organs that generally do not produce them. Plants have evolved secondary metabolites with a broad spectrum of biochemical and pharmacological properties. Currently, many plant secondary metabolites are well known for their diverse biological activities. Broadly, these activities include antibacterial, antifungal and anticancer activities. Around 20,000 plant species are used in traditional medicines and they are prospective reservoirs for new pharmaceuticals. In recent years, traditional medicinal plants have received huge attention as their phytochemicals may lead to new drug discoveries [1]. It is now well established that different plants and their various parts possess different phytocompounds.

### 10.2.1 Plant secondary metabolites

Recently, focus has been on utilizing ecofriendly and biological safe plant-derived products of industrial demand or repute. The approach may be to increase nutrients or to prevent and treat the various diseases. Plants produce thousands of different types of chemicals, including pharmaceutical components, aromatic compounds, pigments, cosmetics, nutraceuticals and other chemicals of industrial importance (Figure 10.1). Plant cells carry out both primary and secondary metabolism [2]. Primary metabolism involves the synthesis of core polymers, especially polysaccharides, proteins, lipids, nucleic acids and amino acids. Secondary metabolism is particularly active at specific stages of growth and development. The environmental factors and microbial interactions may also be responsible for inducing the secondary metabolism in plants at a specific stage or season. Generally, secondary metabolites are derived from primary metabolites through chemical modifications like methylation, hydroxylation and glycosidation. Therefore, secondary metabolites are more complex in

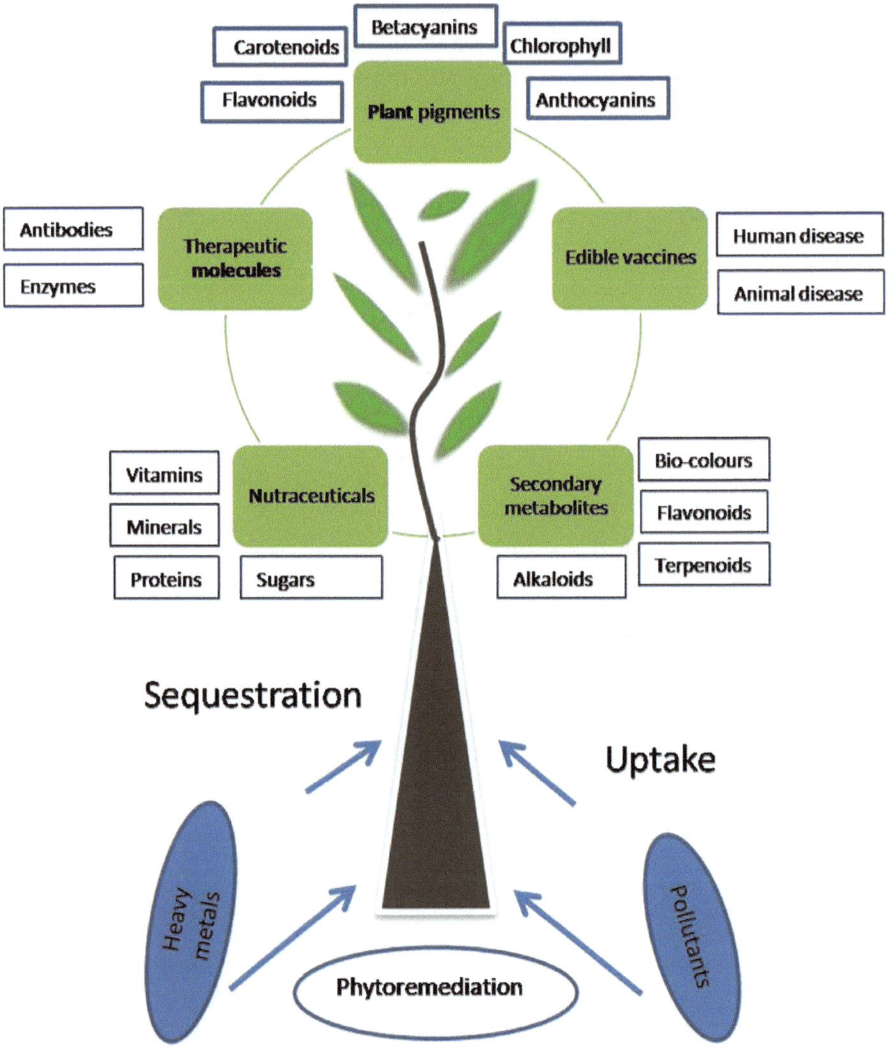


Figure 10.1: Industrial aspects of plant biotechnology.

comparison to primary metabolites [2, 3]. Secondary metabolites are classified on the basis of their chemical structure (aromatic rings and sugars), composition (presence or absence of nitrogen content) and solubility in various solvents. Some secondary metabolites are classified according to the pathways by which they are synthesized. Some of the basic and important secondary metabolites are categorized as alkaloids, phenolics, glycosides, flavonones, flavonoids and flovonols, terpenes and saponins.

## 10.2.2 Phytomedicines

Phytomedicines are plant-based therapeutic molecules or formulations having various active components. Nature is one of the most powerful and primitive sources of drug that include plants, animals and microbes and mineral sources. The drugs derived from the Nature are called natural products. Natural products, especially phytochemicals, are useful in search of novel chemicals entities. Scientifically, the routine procedure involves various steps to develop a commercial formulation or prepare the phytomedicine. These steps are to be followed sequentially to successfully employ the phytomedicines in market. The various steps involved are as follows:
1. Collection of plant material
2. Drying and powder formation
3. Dissolve in solvents for extraction
4. Filtration to get plant filtrate/extract
5. Screening for bioactivity
6. Phytochemical analysis
7. Purification of potent phytochemicals
8. Characterization of most effective bioactive phytochemical agent
9. In vivo studies of most effective bioactive phytochemical agent
10. Final clinical trials and FDA approval
11. Packaging and selling

The phytomedicines have some oblivious advantages that make them molecules of interest in the modern era of drug design development. The natural products are still contributing a major part in the form of starting material for drug discovery. A large number of new chemical entities are either natural derivatives or semisynthetic derivatives of natural products. The plant-based metabolites have been shown to have antimicrobial, antidiabetic, anticancer, antihypertensive and anti-inflammatory properties.

### 10.2.2.1 Phytomedicines for cardiovascular diseases

Cardiovascular diseases (CVDs) are among the leading causes of death globally [4, 5]. Herbal medicine contains various biologically active natural products. A number of herbal products have been employed at global level for the management of CVD [5, 6]. Moreover, this culture has been passed on to the modern generation [5, 6]. Intake of certain flavonoid classes reduces cardiovascular disease risk [7]. Some experts have recommended increased intake of flavonoid-rich diets as preventive measure [7]. Also, consumption of plant foods is associated with lower risk of CVD [8, 9].Flavonoids are bioactive noncaloric, polyphenolic and non-nutrient secondary metabolites in plants. Flavonoids have potent antioxidant, anti-inflammatory activity, reduce LDL cholesterol oxidation, regulate endothelial nitric oxide synthesis and some also have weak estrogenic activity [10]. Phenolics are derivatives of benzene (cyclic derivatives in the case of polyphenols) with one or more hydroxyl groups associated with ring structure. They are classified into 10 different groups, depending upon their chemical structure. The two main types of polyphenols are flavonoids and phenolic acids. Flavonoids are distributed as flavones, flavanones, isoflavones, flavonols, flavanols, proanthocyanidins and anthocyanins. The common flavonoids are quercetin (a flavonolpresent in onion, tea and apple), catechin (a flavanolpresentin tea and several fruits), hesperetin (a flavanone present in citrus fruits), pigment anthocyanin (red fruits; tomatoes, blackcurrant, watermelon, raspberry, pomegranate, strawberry), daidzein (isoflavone in soybean) and proanthocyanidins (commonly present in apple, grape and cocoa). Guggul (*Commiphora wightii*), Garlic (*Allium sativum*), Arjuna (*Terminalia arjuna*) and Hawthorn (*Crataegus oxyacantha*) have been reported for their cardiovascular effects. These plants are known for use in the treatment of heart disease for hundreds of years and can be utilized effectively in the treatment of cardiovascular diseases, including ischemic heart disease, congestive heart failure, arrhythmias and hypertension [11].

#### 10.2.2.2 Phytomedicines as antioxidants

Oxidative stress has been identified as the major reason behind the development and progression of numerous diseases [12]. Plants have an innate ability to synthesize a wide range of antioxidants (nonenzymatic) capable of attenuating reactive oxygen species (ROS)-induced oxidative damage. Although monoterpenes, sesquiterpenes, diterpenes and other alkaloids have shown antioxidant activity, phenolic antioxidants appear to be the most important due to their promising antioxidant activity under both *in vivo* and *in vitro* conditions. Plant phenolics are mainly classified into five major groups: phenolic acids, flavonoids, lignans, stilbenes and tannins [12–15]. Flavonoids and phenolic acids have been found to have excellent antioxidant activity in both *in vitro* and *in vivo* studies. They are also known to interact with other physiological antioxidants, including ascorbate or tocopherol, and to synergistically amplify their biological effects [16]. It can be concluded that phytochemicals play a vital role in preventing oxidative damage. At cellular and molecular levels, they inactivate ROS. Natural antioxidant phytochemicals in food are of great interest due to their beneficial effects on human health. These antioxidants offer protection against oxidative deterioration [17].

#### 10.2.2.3 Phytomedicines against diabetes

Diabetes mellitus is a complex metabolic disorder resulting from insulin insufficiency or insulin dysfunction. Type I diabetes (insulin-dependent) is caused due to insulin insufficiency. Type II diabetes (insulin-independent) is due to immunological destruction of pancreatic β-cells that results in insulin deficiency. Type II diabetes is characterized with insulin resistance. It is the more common form of diabetes constituting major part of the diabetic population [18]. Since ancient times, traditional herbal medicines are in use all over the world. The side effects of insulin therapy and other oral hypoglycemic agents impose the use of more effective and safer antidiabetic drugs. Several medicinal plants have shown antidiabetic activity in experimental and clinical investigations. Plants such as *Momordica charantia, Mucuna pruriens, Murraya koenigii, Eugenia jambolana* and *Brassica juncea* have been reported to have antidiabetic properties [19]. In the recent years, hundreds of plant-derived medicines have been reported to cure and treat diabetic. Major attention has been given to *Vernonia amygdalina, Morinda lucida, Carica papaya, Citrus aurantiifolia, Bidens pilosa, Ocimum gratissimum* and *M. charantia*. Besides these herbal candidates, aloe vera has also been reported to have beneficial effect in treatment of diabetes [20]. The clinical trials of aloe vera have resulted in positive results. The oral administration of aloe vera juice resulted in reduced blood sugar and triglyceride levels in the treated patients. The results suggested the potential of aloe vera juice for use as an antidiabetic agent. The positive effects of aloe vera are due to the presence of compounds such as polysaccharides, mannans, lectins and anthraquinines [21].

### 10.2.3 Plant resources as nutraceuticals

As already mentioned, plants have been used as food and medicine since ancient times and they are important part of our routine diet [22]. The term "nutraceutical" may be defined as food or part of food that provides medical or health benefits, including the prevention and/or treatment of diseases [23, 24]. It can also be defined as diet supplement that delivers a concentrated form of a biologically active component of food in a nonfood matrix to enhance health [25, 26]. It also refers to natural functional or medical food or bioactive phytochemicals that have health-promoting potential and also disease-preventing medicinal properties. Nowadays, a large group of consumers recognize the relationship between nutrition and health, in relation to "functional foods" and "nutraceuticals" [22]. Therefore, the demand for functional food and nutraceuticals is growing rapidly and the associated market is growing sharply. Consumers are now aware of diet-related health problems and need to increase consumption of fruits and vegetables in daily routine.

According to FAO [27], phytochemicals and their metabolic products may inhibit pathogenic bacteria while stimulating the growth of beneficial bacteria, exerting prebiotic and symbiotic (probiotic and prebiotic) effects. Interaction between functional food components, such as prebiotic, probiotic, phytochemicals and intestinal microbiota, have consequences on human health. There are various plant sources of nutraceuticals, including vegetables, fruits, seeds and spices available in nature. A balanced intake of vegetables, beans, fruit and grains can provide an array of beneficial compounds. Studies have demonstrated multiple effects of vegetables and fruits, for example, low concentration of fat, salt and sugar. Moreover, they are good source of dietary fiber. They contain vitamins and minerals good for health, including vitamins A (β-carotene), C and E, magnesium, folic acid, phosphorous and zinc [26]. Plants are rich sources of alkaloids, flavonoids and other beneficial components that can be specified as part of functional foods and nutraceuticals. It may be possible to alter the metabolic pathways and mechanisms to specifically enhance the production of desired metabolites. Various plant parts can be used on daily basis to serve as nutraceuticals and functional foods. Plants-based nutraceuticals and functional foods are safe to consume. Common examples of plant nutraceuticals include glucosamine from ginseng, Omega-3 fatty acids from linseed, *Epigallocatechin gallate* from green tea and lycopene from tomato [28]. Currently, scientists are working on nutraceuticals that can also be used for food preservation purposes.

## 10.3 Phytoremediation

Phytoremediation may be defined as "application of plants for *in situ*, or in place, removal, degradation, or containment of contaminants in soil, sludge, sediments, surface water and groundwater." It is very useful at sites with shallow and low levels of contamination. Phytoremediation is particularly useful for treating a wide variety of environmental contaminants.

The following plant-mediated processes are useful for solving environmental problems:

1. *Phytosequestration* – It is an environment-friendly approach that uses plants to clean up soil from trace element contamination. It may be defined as the ability of plants to sequester certain contaminants in root zone. Various aquatic plants (*Marsilea quadrifolia, Hydrilla verticillata* and *Ipomea aquatica*) as well as terrestrial plants (*Sesbania cannabina* and *Eclipta alba*) have been found efficient accumulator of metals [29].
2. *Phytoextraction* – It involves uptake and concentration of matter from the environment into the plant biomass.
3. *Phytostabilization* – It is the ability to reduce the mobility of specific substances in the environment, for example, by limiting the leaching of substances from the soil.
4. *Phytotransformation* – It is the chemical modification of environmental substances due to plant metabolism, which often results in their degradation (phytodegradation), inactivation or immobilization (phytostabilization).
5. *Phytostimulation* – It involves the enhancement of microbial activity of soil for degradation of contaminants by organisms that are associated with roots. This

process is also known as rhizosphere degradation. Phytostimulation also involves aquatic plants supporting the active populations of microbial degraders, as in the stimulation of atrazine degradation by hornwort [30].

6. *Phytovolatilization* – It is the process of removal of substances from soil or water and release into the air. Sometimes it is due to the phytotransformation to more volatile and/or less polluting substances.
7. *Rhizofiltration* – It is the process of water filtration through a mass of roots to remove toxic substances or excess nutrients. The pollutants remain absorbed in roots. Biological hydraulic containment through the plant decreases the movement of soluble contaminants deeper into the site and into the groundwater [31].

## 10.4 Plant molecular farming for industrial enzymes and therapeutic proteins

Industrial enzymes are useful for food, feed, textile, paper, detergents, meat and energy industries. Various studies have demonstrated the feasibility of transgenic plants to accumulate industrial enzymes (alpha amylase, laccase, phytase) [32]. The heterologous gene expression is a good method to develop new pathways in plant and animals. The heterologous expression of α-amylase from *Bacillus licheniformis* was reported for the first time in tobacco plant [33]. This enzyme is commonly used in the food industry to produce high-fructose corn syrup, ethanol or detergents to remove starch [34]. The enzyme transglutaminase commonly used in the food industry has also been expressed in transgenic rice [35]. Glucosyltransferases (gtfD) expression from *Streptococcus mutans* in transgenic maize seed led to the accumulation of novel glucans, polymers with many applications in food industry, for example, as thickening agent [36]. In the last 30 years, transgenic plants have been predominantly utilized as a factory for the production of economically valuable recombinant proteins (also known as molecular farming). Several examples of molecular farming products are industrial enzymes, antibodies, antigens and vaccines [37, 38]. Plant system offers several advantages in terms of scalability, relatively low cost and free from human or animal pathogens, thus providing a safe environment [39]. Human growth hormone and interferon were among the first molecular farming products that have been successfully expressed in plants [40, 41]. Several plants-derived recombinant pharmaceuticals such as avidin [42–44] and beta glucuronidase [45] have been produced and commercialized.

The various industrial enzymes and their applications produced in plant tissue culture and transgenic plants are amylases (food processing, paper industry), avidin and beta-glucuronidase (diagnostic kits), cellulases (ethanol and paper industry, dish washing detergents), glucanase (brewing industries), lignin peroxidase (paper industry), trypsin (pharmaceutical industry), pepsin (cheese industry), xylanses (biomass processing, paper and pulp industry) and phytase (better phosphate utilization).

Plants have been used to manufacture drugs, antibodies and vaccines. One of the biopharmaceuticals, enkephalins, expressed in transgenic plant is useful as pain killer [46]. FDA has approved a drug taliglucerase alpha expressed by transgenic carrot cell culture for Gaucher disease treatment [47].

Recently, plant-derived products have been approved for commercialization or are in the clinical trials stage [39, 48, 49]. Plant derived antibody has been reported by researchers [50]. Accumulation and assembly of immunoglobulins in tobacco has been carried out successfully. More evidences regarding plant producing antibodies were reported afterward [39, 51, 52]. Arabidopsis-derived antibodies (IgA) designed for the passive immunization against enterotoxigenic *Escherichia coli* infection prolonged protection of weaned piglets from diarrhea [53]. In 2014, the plant antibodies were derived from tobacco to treat Ebola virus in human [54]. In order to prevent diseases, vaccines have to be quickly produced in enough quantity. Transient expression in plants offers production system of pharmaceuticals that can be achieved in 2–4 weeks [55]. Plant-derived vaccines also offer oral delivery that possibly removes the necessity of needle injection and the requirement of protein purification, and thus allows easy administration [56]. Oral administration also enhances and induces mucosal immunity, which is an important factor for stronger immune response. Plant-derived vaccines have been produced for pathogenic diseases such as hepatitis B, enterotoxin *E.coli* and influenza [57–59]. According to Sandhu et al. [60], orally fed mice with transgenic tomato expressing respiratory syncytial virus-F protein (RSV-F) stimulate the production of IgG and IgA antibodies indicating serum and mucosal immunity.

## 10.5 Conclusion

The biotechnological techniques such as rDNA technology and metabolic engineering can be useful for altering or enhancing synthesis of various industrial important molecules and metabolites. The plant metabolites are safe to consume as compared to their chemical counterparts. Besides therapeutics, functional foods and nutraceuticals, the plants resources can also be used for environmental issues. Plant resources can also be used as production sites for pigments, colors, alkaloids, flavonoids, therapeutic agents, edible vaccines and phytoremediation. In coming years, more population will move toward the plants-based products and this will increase the demand of phytoproducts. Immense research and commercialization is required to satisfy the demand of plant products having no side effects, safe to consume and simultaneously avoiding social issues.

**Acknowledgement:** The authors (RKP, PJ, SK, RK) are thankful to management of Ambala College of Engineering and Applied Research (ACE) Ambala, Haryana for encouraging us to write this informative manuscript.

# References


[1] Amor ILB, Boubaker J, Sgaier MB, Skandrani I, Bhouri W, Neffati A, Kilani S, Bouhlel I, Ghedira K, Chekir-Ghedira L. Phytochemistry and biological activities of *Phlomis* species. J. Ethnopharmacol 2009, 125, 183–202.

[2] Sharma S, Shahzad A. Sahai A. Hairy root culture: an efficient system for secondary metabolite production. In: Shahid M, Shahzad A, Malik A, Sahai A., ed. Recent trends in biotechnology and therapeutic applications of medicinal plants. Springer, The Netherlands, 2013, pp 51–78.

[3] Halliwell B, Whiteman M. Measuring reactive species and oxidative damage in vivo and in cell culture: How should you do it and what do the results mean? Br J Pharmacol. 2004, 142, 231–255.

[4] Zheng CS, Xui XJ, Yei HZ, Wu CW, Xu HF, Li XH, Huang SP, Liu XX. 2013. Computational pharmacological comparison of *Salvia miltiorrhiza* and *Panaxnotoginseng* used in the therapy of cardiovascular diseases. ExpTher Med 2013, 6, 1163–1168.

[5] Suroowan S, Mahomoodally F. Common phyto-remedies used against cardiovascular diseases and their potential to induce adverse events in cardiovascular patients. Clinical Phytoscience 2015, Springer Berlin Heidelberg, 1:1. https://doi.org/10.1186/s40816-015-0002-3.

[6] Lin MC, Nahin R, Gershwin ME, Longhurst JC, Wu KK. State of complementary and alternative medicine in cardiovascular, lung, and blood research: executive summary of a workshop. Circulation, 2001, 103, 2038–2041.

[7] Peterson JJ, Dwyer JT, Jacques PF, McCullough ML. Do flavonoids reduce cardiovascular disease incidence or mortality in US and European populations? Nutr Rev 2012, 70, 491–508.

[8] Dauchet L, Amouyel P, Dallongeville J. Fruits, vegetables and coronary heart disease. Nat Rev Cardiol 2009, 6, 599–608.

[9] McCullough ML, Peterson JJ, Patel R, Jacques PF, Shah R, Dwyer JT. 2012. Flavonoid intake and cardiovascular disease mortality in a prospective cohort of US adults. Am J ClinNutr 2012, 95, 454–464.

[10] Middleton E, Kandaswami C, Theoharides TC. The effects of plant flavonoids on mammalian cells: implications for inflammation, heart disease, and cancer. Pharmacol Rev 2000, 52, 673–751.

[11] Rastogi S, Pandey MM, Rawat A.K. Traditional herbs: a remedy for cardiovascular disorders. Phytomed 2016, 23, 1082–1089.

[12] Kasote DM, Katyare SS, Hegde MV, Bae, H. Significance of Antioxidant Potential of Plants and its Relevance to Therapeutic Applications. Int J Biol Sci. 2015, 11, 982–991.

[13] Duthie GG, Duthie SJ, Kyle JA. Plant polyphenols in cancer and heart disease: implications as nutritional antioxidants. Nutr Res Rev 2000;13, 79–106.

[14] Myburgh KH. Polyphenol supplementation: benefits for exercise performance or oxidative stress? Sports Med. 2014, 44, S1, S57–70.

[15] Blokhina O, Virolainen E, Fagerstedt KV. Antioxidants, oxidative damage and oxygen deprivation stress: a review. Ann Bot 2003, 91, 179–94.

[16] Croft KD. The chemistry and biological effects of flavonoids and phenolic acids. Ann N Y AcadSci 1998, 854, 435–442.

[17] Girija K, Radha R. Role of phytomedicines and neutraceuticals in the prevention of chronic diseases. Inter J Pharm Sci Res. 2013, 4, 3363–3368.

[18] Grover JK, Yadav S, Vats V. Medicinal plants of India with antidiabetic potential. J Ethnopharmacol 2002, 81, 81–100.

[19] Chandrakar M, Palekar S, Chirade SG, Almas S, Hafiz M. Hypocholesterolemic effect of *Aloe vera* (L.) extract on high cholesterol fed calotes versicolor daudin. Asian J. Exp. Sci 2008, 22, 295–298.

[20] Abo-Youssef AMH, Messiha BAS. Beneficial effects of *Aloe vera* in treatment of diabetes: Comparative *in vivo* and *in vitro* studies. Bulletin of Faculty of Pharmacy, Cairo University 2013, 51, 7–11.
[21] Yazaki K, Sugiyama A, Morita M, Shitan N. Secondary transport as an efficient membrane transport mechanism for plant secondary metabolites. Phytochem Rev 2008, 7, 513–524.
[22] Pinela J, Carocho M, Dias MI, Caleja C, Barros L, Ferreira ICFR. Wild Plant-Based Functional Foods, Drugs, and Nutraceuticals. In: ICFR. Ferreira, Morales P, Barros L. Wild Plants, Mushrooms and Nuts: Functional Food Properties and Applications. 2016, ISBN: 9781118944622.
[23] Keservani RK, Kesharwani RK, Vyas N, Jain S, Raghuvanshi R, Sharma AK. 2010. Nutraceutical and functional foods as future food: A review. Der Pharm Lett 2010, 2, 106–116.
[24] Jain P, Pundir RK. Nutraeuticals: Recent Developments and Future prospectives. In: Shahid M, Shahzad A, Malik A, Sahai A (eds.), Recent trends in Biotechnology and therapeutic applications of medicinal plants. 2013, 213–224, Print ISBN 978-94-007-6602-0.
[25] Zeisel SH. Regulation of "nutraceuticals". Science 1999, 285, 1853–1855.
[26] Varzakas T, Zakynthinos G, Verpoort F. Plant Food Residues as a Source of Nutraceuticals and Functional Foods. Foods, 2016, 5, E88.
[27] Food and agriculture organization of the united nation (FAO). Report on functional foods, food quality and standards service (AGNS). 2007, Rome.
[28] Pandey N, Meena RP, Rai SK, Rai SP. Medicinal plants derived nutraceuticals : a re-emerging health aid. Inter J Pharm Bio Sci 2011, 2, 419–441.
[29] Anwar A, Ghufran R, Zularisam AW. Phytosequestration of metals in selected plants growing on a contaminated okhla industrial areas, Okhla, New Delhi, India. Water, Air, & Soil Pollution 2011, 217, 255–266.
[30] Rupassara SI, Larson RA, Sims GK, Marley KA. Degradation of atrazine by hornwort in aquatic systems. Bioremediat J 2002, 6, 217–224.
[31] Evans GM, Furlong JC. (2010). Phytotechnology and Photosynthesis. In: Environmental Biotechnology: Theory and Application. John Wiley & Sons, Ltd. pp. 145–174. ISBN: 9780470684184.
[32] Howard JA, Hood EE. Strategies to maximize recombinant protein expression in maize kernels. In: Azhakanandam K, Silverstone A, Daniell H, Davey MR (eds) Recent Advancements in Gene Expression and Enabling Technologies in Crop Plants. Springer Publishers, New York, 2015, 79–129.
[33] Pen J, Molendijk L, Quax WJ, Sijmons PC, van Ooyen AJ, van den Elzen PJ, Rietveld K, Hoekema A. Production of active *Bacillus licheniformis* alpha-amylase in tobacco and its application in starch liquefaction. Biotech NY 1992, 10, 292–296.
[34] Richardson TH, Tan X, Frey G, Callen W, Cabell M, Lam D, Macomber J, Short JM, Robertson DE, Miller C. A novel, high performance enzyme for starch liquefaction. Discovery and optimization of a low pH, thermostable alpha-amylase. J BiolChem 2002, 277, 26501–26507.
[35] Claparols MI, Bassie L, Miro B, Del Duca S, Rodriguez-Montesinos J, Christou P, Serafini-Fracassini D, Capell T. (2004) Transgenic rice as a vehicle for the production of the industrial enzyme transglutaminase. Transgenic Res 2004, 13, 195–199.
[36] Zhang S, Dong JG, Wang T, Guo S, Glassman K, Ranch J, Nichols SE. High level accumulation of alpha-glucan in maize kernels by expressing the gtfD gene from *Streptococcus mutans*. Transgenic Res 2007, 16, 467–478.
[37] Ramessar K, Peremarti A, Gómez-Galera S, Naqvi S, Moralejo M, Muñoz P, Capell T, Christou P. Biosafety and risk assessment framework for selectable marker genes in transgenic crop plants: a case of the science not supporting the politics. Transgenic Res 2007, 16, 261–280
[38] Sack M, Hofbauer A, Fischer R, Stoger E. The increasing value of plant-made proteins. CurrOpinBiotechnol 2015, 32, 163–170.

[39] Twyman RM, Stoger E, Schillberg S, Christou P, Fischer R. Molecular farming in plants: host systems and expression technology. Trends Biotechnol 2003, 21, 570–578.
[40] Barta A, Sommergruber K, Thompson D, Hartmuth K, Matzke MA, Matzke AJ. The expression of a nopaline synthase – human growth hormone chimaeric gene in transformed tobacco and sunflower callus tissue. Plant MolBiol 1986, 6, 347–357.
[41] De Zoeten GA, Penswick JR, Horisberger MA, Ahl P, Schultze M, Hohn T. The expression, localization, and effect of a human interferon in plants. Virology 1989, 172, 213–222.
[42] Hood EE, Witcher DR, Maddock S, Meyer T, Baszczynski C, Bailey M, Flynn P, Register J, Marshall L, Bond D, Kulisek E, Kusnadi A, Evangelista R, Nikolov Z, Wooge C, Mehigh RJ, Hernan R, Kappel WK, Ritland D, Li CP, Howard JA. Commercial production of avidin from transgenic maize: characterization of transformant, production, processing, extraction and purification. Mol Breed 1997, 3, 291–306.
[43] Hood EE. From green plants to industrial enzymes. Enzyme MicrobTechnol 2002, 30, 279–283.
[44] Woodard SL, Mayor JM, Bailey MR, Barker DK, Love RT, Lane JR, Delaney DE, McComas-Wagner JM, Mallubhotla HD, Hood EE, Dangott LJ, Tichy SE, Howard JA. Maize (Zea mays)-derived bovine trypsin: characterization of the first large-scale, commercial protein product from transgenic plants. BiotechnolApplBiochem 2003, 38, 123–130.
[45] Witcher DR, Hood EE, Peterson D, Bailey M, Bond D, Kusnadi A, Evangelista R, Nikolov Z, Wooge C, Mehigh R, Kappel W, Register J, Howard JA. Commercial production of β-glucuronidase (GUS): a model system for the production of proteins in plants. Mol Breed 1998, 4, 301–312.
[46] Maxmen A. Drug-making plant blooms. Nature, 2012, 485, 160.
[47] Vandekerckhove J, Damme JV, Lijsebettens MV, Botterman J, Block MD, Vandewiele M, Clercq AD, Leemans J, Montagu MV, Krebbers E. Enkephalins prouduced in transgenic plants using modified 2S seed storage proteins. Nat Biotechnol 1989, 7, 929–932
[48] Fahad S, Khan FA, Pandupuspitasari NS, Ahmed MM, Liao YC, Waheed MT, Sameeullah M, Darkhshan, Hussain S, Saud S, Hassan S, Jan A, Jan MT, Wu C, Chun MX, Huang J. Recent developments in therapeutic protein expression technologies in plants. BiotechnolLett, 2015, 37, 265–279.
[49] Moustafa K, Makhzoum A, Trémouillaux-Guiller J. Molecular farming on rescue of pharma industry for next generations. Crit Rev Biotechnol 2016, 36, 840–850.
[50] Hiatt A, Cafferkey R, Bowdish K. Production of antibodies in transgenic plants. Nature 1989, 342, 76–78.
[51] Artsaenko O, Kettig B, Fiedler U, Conrad U, Düring K. Potato tubers as a biofactory for recombinant antibodies. Mol Breed 1998, 4, 313–319.
[52] Fischer R, Schumann D, Zimmermann S, Drossard J, Sack M, Schillberg S. Expression and characterization of bispecific single-chain Fv fragments produced in transgenic plants. Eur J Biochem 1999, 262, 810–816.
[53] Virdi V, Coddens A, De Buck S, Millet S, Goddeeris BM, Cox E, De Greve H, Depicker A. Orally fed seeds producing designer IgAs protect weaned piglets against enterotoxigenic *Escherichia coli* infection. ProcNatlAcadSci USA 2013, 110, 11809–11814.
[54] Qiu X, Wong G, Audet J, Bello A, Fernando L, Alimonti JB, Fausther-Bovendo H, Wei H, Aviles J, Hiatt E, Johnson A, Morton J, Swope K, Bohorov O, Bohorova N, Goodman C, Kim D, Pauly MH, Velasco J, Pettitt J, Olinger GG, Whaley K, Xu B, Strong JE, Zeitlin L, Kobinger GP. Reversion of advanced Ebola virus disease in nonhuman primates with ZMapp Nature 2014, 514, 47–53.
[55] Negrouk V, Eisner G, Lee H-i, Han K, Taylor D, Wong HC (2005) Highly efficient transient expression of functional recombinant antibodies in lettuce. Plant Sci 169, 433–438.
[56] Nochi T, Takagi H, Yuki Y, Yang L, Masumura T, Mejima M, Nakanishi U, Matsumura A, Uozumi A, Hiroi T, Morita S, Tanaka K, Takaiwa F, Kiyono H. Rice-based mucosal vaccine as a global

strategy for cold-chain- and needle-free vaccination. ProcNatlAcadSci USA 2007, 104, 10986–10991.

[57] Kapusta J, Modelska A, Figlerowicz M, Pniewski T, Letellier M, Lisowa O, Yusibov V, Koprowski H, Plucienniczak A, Legocki AB. A plant-derived edible vaccine against hepatitis B virus. FASEB J. 1999, 13, 1796–1799.

[58] Chikwamba R, Cunnick J, Hathaway D, McMurray J, Mason H, Wang K. A functional antigen in a practical crop: LT-B producing maize protects mice against *Escherichia coli* heat labile enterotoxin (LT) and Cholera Toxin (CT). Transgenic Res 2002, 11, 479–493.

[59] Nahampun HN, Bosworth B, Cunnick J, Mogler M, Wang K. (2015) Expression of H3N2 nucleoprotein in maize seeds and immunogenicity in mice. Plant Cell Rep 2015, 34, 969–980.

[60] Sandhu JS, Krasnyanski SF, Domier LL, Korban SS, Osadjan MD, Buetow DE. Oral immunization of mice with transgenic tomato fruit expressing respiratory syncytial virus-F protein induces a systemic immune response. Transgenic Res 2000, 9, 127–135.

# Index



https://doi.org/10.1515/9783110563337-011